HOW TO BUILD IMPOSSIBLE THINGS

'A brilliantly engaging storyteller, laugh-out-loud funny, loving, cheekily smug . . . An enjoyable read on making, inventing and what might contribute to a life worth living.'
Julie Mehretu

'Mark is an amazing polymath – and an Olympic-level aesthete. Unlike many polymaths and aesthetes, though, when he gets up in the morning, it's to make real, physical things – including this book.'
Craig Nevill-Manning, Engineering Director, Google NYC

'On a job site Mark makes irreverent banter while scribbling measurements on the back of a pizza box as works of astonishing complexity and precision materialize under his direction. Now he has somehow applied this same deceptively offhand but exacting craft to unspooling this collection of tales from his ascent to the summit of one of the most demanding construction habitats on earth.'
David Hotson, architect, Skyhouse and Pinnacle

Mark Ellison is regarded by many as the best carpenter in New York. A man with an affinity for challenging work, he has designed and constructed some of New York's most elaborate and expensive homes, and been profiled in the *New Yorker*. But, as a native of the old steel town Pittsburgh, Pennsylvania, his route into the building trade and the mastery of a craft was unexpected, moving from construction labourer to helper and finally to carpenter. Now, at the age of sixty, he has written his first book.

HOW TO BUILD IMPOSSIBLE THINGS

Lessons in Life and Carpentry

MARK ELLISON

PENGUIN BOOKS

PENGUIN BOOKS

UK | USA | Canada | Ireland | Australia
India | New Zealand | South Africa

Penguin Books is part of the Penguin Random House group of companies
whose addresses can be found at global.penguinrandomhouse.com

First published in the US by Random House, an imprint and division
of Penguin Random House LLC, New York, in 2023
First published in the UK by Hutchinson Heinemann in 2023
Published in Penguin Books 2024
001

Book design by Caroline Cunningham

Printed and bound in Great Britain by Clays Ltd, Elcograf S.p.A.

The authorised representative in the EEA is Penguin Random House Ireland,
Morrison Chambers, 32 Nassau Street, Dublin D02 YH68

A CIP catalogue record for this book is available from the British Library

ISBN: 978–1–529–15730–7

www.greenpenguin.co.uk

For Pauline Kow. Okay, I'll write it.

CONTENTS

An Inauspicious Beginning

I am a carpenter. I build things.

I didn't set out to become a carpenter. The trade found me by chance. After leaving high school for the last time, I was offered work in Cambridge's Central Square fixing up the newly acquired townhouse of a classmate's parents. Repairs in my childhood home were done by someone called Dad, and anyone who could help was expected to.

As a child, it never occurred to me, outside of Sunday school, that anyone had an occupation called carpenter. But given a chance at the work, I took to it immediately; I was handy, unafraid of grimy labor, and took pride in the job title I hadn't earned yet.

There was a time in my career when I would lie about my work experience. At thirty, I would say that I had been a carpenter, a cabinetmaker, or whatever I was calling myself that day for more than fifteen years. I was young for a craftsman and hoped to steal some old-timer credibility. Of course, the math didn't add up, but no one was checking IDs, and even back then, carpenters with the ten years of experience I actually had weren't so easy to come by.

There's no need to lie anymore; it really has been forty years. I still love building. I always have; that's how I've managed to keep at it for so long. Every day, I find it technically, interpersonally, and at times physically and psychologically challenging. In the past, I would lie awake nights wishing that I'd get to build a home so splendid that it rivaled the great mansions of the Gilded Age, or so space-age modern the Jetsons would be jealous. As the years came and went, those wishes and many more came true.

Most of the places I have built in the last thirty years are so opulent or sleek that they resemble the homes I've inhabited in name only. They are to shelter what the Taj Mahal is to tombstones. The cleverest of New York's design community know full well that the biggest commissions are to be had at the point where sumptuousness and singularity collide. I have had more than one client who has openly reveled in the thought that they were the only person alive who would possess the room in which they stood. The most extravagant places I've built have an unreality to them, even for me. Thinking back, I marvel that I was even allowed through the door.

I've never had a baby, but I've seen it done a few times, and I've heard it said that if women could remember the effort and pain of childbirth the species would die out; no one would ever do *that* twice by choice.

Everyone loves a good baby picture; they're cute when they're cleaned up and quiet. Likewise, everyone loves a good interior design photo spread—the perfect details, the wonderfully rich materials, the light glancing off every surface just so—all cleaned up for the camera. I can never look at pictures of fancy New York apartments in the envious oohing-and-aahing way. I know all too well that everything, in every shot, was unloaded from a truck on the street, maybe in the rain or snow, wedged into a freight elevator, and painstakingly modified onsite to fit seamlessly into a dust-choked, crooked space ten times the size of anything I have ever called home, with neighbors on all sides who are practiced at complaining loudly, even litigiously, that the noisemaking oafs next door should be silenced, or better yet, imprisoned.

Every glossy interior design magazine is a baby album to someone like me. I am proud of all those pictures, I look back at the best of them from time to time and reminisce, but the serenity these too-pretty pictures project always strikes me as absurd, given the chaos that brought them into being. Where are the grime-covered workers, the 110-degree days, the clients' screaming fits, the injuries, the career-ending errors?

For the most part, all we see of the world is its reflective surface. From my window in Long Island City, I often watch Manhattan's skyline make its passage from sharp-edged morning to glowing, blending dusk. I can see a few of the buildings I've worked in and picture the finished apartments. There is little difference between my high-rise middle-income flat and those exclusive towers. The materials that go into their making are identical: structural concrete and steel, plate glass windows in aluminum frames, metal studs with drywall finishes, wood composite doors, and whatever smatterings of hardware, fixtures, and appliances make the place usable. Lights, stoves, doorknobs, and air conditioners fulfill exactly the same functions in fancy apartments as they do in mine—they just cost ten times as much.

The architectural elements of each place are likewise the same; they're just more artfully arranged in a high-end home. The real difference between my apartment and the apartments I build for a living is in their surfaces. Surface makes the show. It's all anyone sees. But the illusion is millimeters thin and can be dispelled with a scratch. The precious wood in fancy millwork these days is rarely more than a sixteenth of an inch thick. Behind that, it's either chewed-up scraps from the mill or the same pedestrian poplar that lines so many small-town streets. The most expensive handmade ceramic tiles are nothing more than a seductive glaze drizzled over refined baked dirt.

In a well-executed renovation, an enormous amount of planning and labor goes into getting every inch of those surfaces just right. I've never had a customer who cares a whit about the systems or the materials that are behind those veneers as long as they keep functioning.

People are willing to pay unthinkable sums for these places because the illusion works. Looking at a photograph of well-coiffed, superbly dressed people in lavishly composed surroundings, few of us can escape envy's grasp. Our entire social order can be measured by envy's accumulated gaze. In this country, and throughout much of the world, we are taught from birth to elevate, emulate, and fairly drool over the richest among us. People bandy about other notions of success—accomplishment, talent, intelligence, wisdom, beauty, and determination—but it is the accumulation of wealth and its trappings that trump all pretenders in the public imagination. My chosen industry, and the enormous publishing machine that drives it, devotes itself entirely to the care and feeding of this myth.

Four decades of building have been an unusual education. My colleagues and I spend almost all of our days behind the surfaces others see. We see everything: the dusty, rubbled skeletons of century-old Park Avenue "prewars"; the dank basements with their shuffling, stooped denizens; the subcontracting crews composed entirely of men whose countries have been overrun by horror; the designers whose sole task is to run up the tab; the harassed housemaids with their hair falling out in clumps; the dressing room drawers stuffed with Adderall and more Ritalin than any precocious prepubescent boy could ever ingest. It has been a long, low education, built with a thousand such stones. Over time, these stones have circled me completely, and finally, envy can no longer enter.

It is beneath the surface that most of life is found. In every carving Michelangelo made, you can see the bones and tendons pressing through the stretching flesh. He dissected cadavers with the same meticulousness with which he cut to the psyche's quick. His discoveries emanate from beneath his polished marbles' surfaces. Every devoted hunter knows that hanging one's kill for weeks and butchering it oneself makes all the difference at the barbecue. If all a person knows is the pretty pictures, the manicured lawns, and the served meals of life, they could miss a lot.

Those of us who build the homes you see in a magazine finish many a day covered in grease and dirt and blood. All the world's advertisers and tin-pot electronic barkers can shout themselves hoarse about the glamour and satisfaction of unbounded accumulation; it can't penetrate our dark armor. We've seen too far behind the veneer. A private palazzo might boost your confidence when the dinner guests swoon, but it can't make your children love you.

I don't think everyone needs to devote themselves to manual labor or a craft, but I firmly believe that most of life passes us by when we avoid mucking about in dirt, or dough, or dark thoughts. Doing anything from beginning to end brings understanding that no finished product can provide. Thirty years ago, I couldn't have said such things; I didn't feel I had much to say at all. Now, at nearly sixty, there are a few things I've learned that I'd like to share.

I do feel compelled to say: This book is not for people who think they want to become fancy carpenters. This book is for people who are interested in doing anything well, hopefully something that *they* want to do, not their parents, nor their teachers, nor anyone else who wears the disapproving scowl of "authority."

I would be the last to say that mastery is an easy road. I won't even use the word "master" in regard to myself, and I get rankled when others try. It's best to acknowledge that in any endeavor, falling short is the rule. Every serious effort is met with resistance, often from unexpected quarters. Who could have predicted that Newton's third law could include events as divergent as losing my last five hundred dollars to a subway pickpocket or having to drive myself to the emergency room to have my wrist sewn back together after nearly severing my left hand with a chain saw?

The laws of this world are not forgiving or supportive; they are concerned with the survival of DNA, not with your fulfillment or mine. But I contend that people are capable of doing a lot more of what they want with their lives than they are led to believe.

The very person who mistook academia as my way was the first to illustrate to me how much an individual can achieve if they really want something unique from life and are willing to put in the work to get it. She is ever an inspiration to me, so I will share her story.

———

My mother is a physician; she *started* medical school when my three siblings and I were of primary school age. It was the late sixties in Pittsburgh. She had to keep us kids a secret at school to escape being branded as delusional.

No one would have predicted the path she chose. By 1956, she had graduated from Cornell University with a degree in botany, one of the few sciences that welcomed women, and was on her way to New York City with my father, where he was to study divinity. While he followed his lofty pursuits, she studied textiles at Teachers College, Columbia University. My brother joined them in their dormitory room in 1958, occupying the lowest drawer of their dresser, as floor space was scarce.

My father passed his exams in a subject about which no living human has any actual knowledge, and they drove across the country so that he could fill his first post as the chaplain of Oregon State Hospital, celebrated in *One Flew Over the Cuckoo's Nest* for its mistreatment of the mentally ill.

Half a year later, my older sister arrived. As life's demands accumulated, my mother informed my father that she was not prepared to spend hers as a preacher's wife.

In 1960, a preacher without a preacher's wife was like a shepherd without a collie. Who was going to do all the actual work? To my father's credit, he took the news to heart and abandoned his burgeoning career. They loaded their secondhand VW microbus and began another slow drive across the country, this time landing in Lafayette, Indiana.

My father pursued a PhD in sociology at Purdue University, a field that at the time appeared to offer a front row seat to society's rapid changes. I didn't want to miss them, so I was born. Eighteen months later, my younger sister arrived to rebalance the brood. With my father's new PhD secured, we all packed up again and moved to Pittsburgh, where he had been offered an assistant professorship.

As was the custom of the era, my mother stayed at home, charged with making family life work while my father found his feet in teaching. Day-to-day life was frugal; I was ten years old before I wore a pair of trousers that were bought for me. By the same age, I had seen two movies in a theater, and I could count on one hand the number of times I had eaten in a restaurant. But my parents were clever; they took advantage of every educational, medical, and recreational perk the University of Pittsburgh had to offer. By arranging lodging in the just-livable second story of a tumbledown barn set beside an Adirondack lake, they even managed ten long summer vacations. This was my childhood Eden.

A dark broad room above the barn became my mother's studio. It was here that she spent her daytime hours sewing the year's au courant additions to her wardrobe. Each sewing season began with a trip to Mrs. Macro's in Tupper Lake. Helen Macro had no rational business doing so, but she would travel to New York each year and bring back that city's most fabulous fabrics— tulle, Ultrasuede, velvet, linen, silk—the best of them, in the bold weaves and prints that the era had proclaimed chic. In Mrs. Macro's basement, in Tupper Lake, in Franklin County, the poorest county in New York State, in plain sight, witchcraft was afoot.

There were, perhaps, five skinny aisles stretching the width of her cellar. At nine, I couldn't walk the length of one without brushing against and breathing in the bolts and swatches perched everywhere. My mother would carry off a dozen yards or so after each visit without leaving any noticeable dent in the inventory.

Only looking back can I surmise from all this that Helen Macro's basement was the underground somatic hub of a vast hidden creative network that to this day has left barely a trace, but whose influential tentacles reach as deeply into our social fabric as my mother's generation of feminists was able to make them penetrate.

A single photograph is the only proof I have of this theory, but it is compelling. Assembled on the hillock of lawn in front of my childhood home, my siblings and I fall off stepwise to my mother's left. Her rosebushes lend the background an elegant air. She is wearing a simple, perfectly fitted, linen A-line shift dress accessorized by a medical school graduation gown. It is the same dress that the summer before she had cut, stitched, zippered, collared, and fit at the oversized table in the room we children never played in, above the creaky old barn.

That day, we had been the only family of six to attend Pitt's medical school commencement ceremony. My mother was the only mother onstage. In 1966, with four children under the age of nine, she had been admitted to the University of Pittsburgh School of Medicine, one year *before* they were successfully sued for holding female applicants to higher standards than males. She graduated at the top of her class. Throughout her school years, I can only recall a handful of evenings when we didn't sit down as a family to eat the dinner she had cooked for us.

We assembled on the lawn shortly after we got home from her ceremony.

This is a photograph of my mother's first day of freedom. It was taken by my father, though he likely still regrets it.

Fifth Avenue fathers of the bride are badgered into spending small fortunes in an attempt to purchase the ersatz version of the metamorphic majesty my mother displayed that day. Anyone who can see deeper than the glossy surface of a magazine's soci-

ety page knows full well it can't be had for hire; real pride is only
for those who have earned it.

Happiest day of a girl's life, indeed.

What I do these days bears little resemblance to what I did forty
years ago. Only about a quarter of my time is spent in my workshop making things. It's not for lack of opportunities; people
propose things all the time that they think I would like to make
for them. After gluing up a thousand or so cabinets with every
imaginable style of door, becoming proficient at the relatively
small number of joints that woodwork requires, and picking up
enough metal, plastics, glass, and mechanical knowledge to
build most things one could find in an elegant home, I am finally able to pick and choose the projects that are interesting—
which usually means challenging—to me. I have had to admit
that I am not really the solitary craftsman type. I miss the jum-
bled, sometimes aggravating chaos of the jobsite when I work
alone in the shop too long.

Most days I'm out in the field unraveling the contradictory
and poorly conceived instructions builders get from architects
and engineers these days. Owners pay me directly for this service, which seems odd, even to me, because they have just paid
their design team for the same service. A day or two with the
drawings they have in hand is enough for me to provide my
clients with some six-figure reasons why it might be prudent to
hire me to intervene. Contractors are contractually obliged to
build what is shown in the drawings. If they don't do as they are
told, they can be sued out of existence. Thankfully, I am not a
contractor. I am hired by nervous owners to clean up messes
that haven't happened yet. My method is to assume that every

document I'm handed is riddled with errors, omissions, over-
sights, and insufficient information. Doing as I am told would
mean building thousands of flaws into every project in which
I'm involved. The thought never enters my mind.

I have never been good at rules. I am terrible at recognizing
authority. Somehow, after all these years of applying myself to
my trade, I have turned these central character flaws into a lucrative and entertaining business model.

Let me take you to my current jobsite—it's a fine example of
how this model works.

Presently, my colleagues and I find ourselves in a pair of side-
by-side wooden townhouses built about 180 years ago in Brooklyn. They have low ceilings, rubble foundations, no insulation,
crumbling finishes, and a fair share of asbestos, mold, and rot.
The first time I walked through them, I could feel the oppres-
sive gloom that living there would surely bring on. If I owned
them, I would pray for their collapse. The Landmarks Preservation Commission finds them charming, so we are required to
preserve their structures and restore the exterior finishes to the
state depicted in a few photographs from the early 1900s, photos
that show clear evidence of at least one layer of extensive stylistic
reworkings. These houses are to historical gems what yard sale
paintings are to fine art, but appreciation is in the eye of the
beholder. With certain authorities, there is no circumvention.
We will "save" the wretched things.

The plans indicate that we are to dig three feet deeper into
the basements to make them habitable. This can't be done without extending the foundations by the same amount. Our engineers proposed that we should use a technique called "A-B-C
underpinning." I've underpinned several projects over the years.
It goes like this:

- Three-foot-wide approach trenches are dug *by hand* from the inside of the basement toward and under the existing foundation. The trenches are spaced six feet apart and are dug around the entire perimeter to whatever depth the foundation is being extended downward. Wooden planks are installed on the trenches' sides to keep them from collapsing as work progresses.
- Concrete and reinforcing steel are placed beneath the existing foundation to the required depth and thickness, and allowed to cure.
- Once the concrete has reached sufficient hardness, the wooden planks are removed and the process begins again by *hand digging* the next complete series of three-foot trenches, reinforcing them with planks, and pouring new concrete beneath those sections of foundation.
- Repeat once more and you're all done!

A·B·C UNDERPINNING

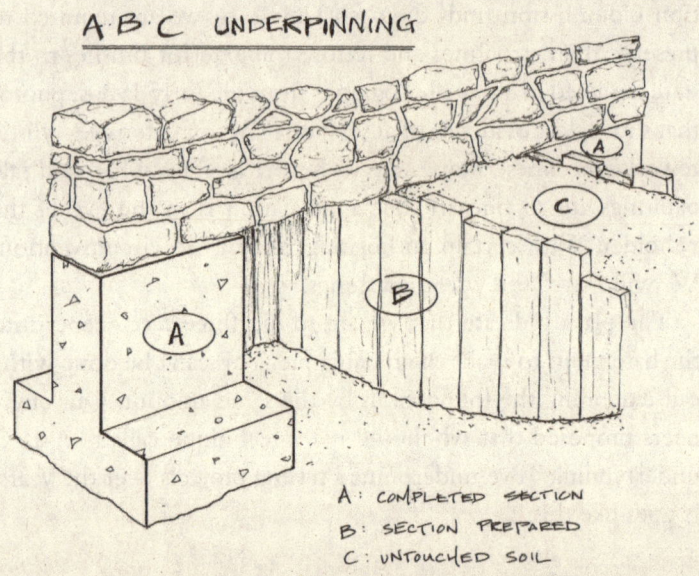

A: COMPLETED SECTION
B: SECTION PREPARED
C: UNTOUCHED SOIL

A-B-C. It sounds like a child could do it. Most children these days haven't done much digging, but I can assure you that performing this process, in hot weather or cold, borders on the inhumane. Even for adults, it is painfully slow, grueling, and wildly expensive work.

This already unpleasant plan graduated to untenable on the first day I visited the houses. Rubble foundations are exactly what they sound like, a pile of inconsistently shaped rocks tossed in a heap with a slather of mortar between them, mortar mixed in the days when underground cementitious slurries had a useful life of a few decades at best. I could grab any stone in the place and pull it out with my bare hands. It doesn't take an engineering degree to imagine what will happen when a three-foot-wide trench is dug beneath a loosely assembled pile of rocks.

We needed another plan.

I often don't have a ready solution for problems of this magnitude. Huge sums of money can be involved; egos get easily bruised; workers' safety is at risk; everyone needs assurance that the new plan is markedly better than the old one. I generally give myself a week or two to empty my head of ideas and mull.

I'm hard-pressed to explain how it happens, but over time, by carrying around a problem with me everywhere I go, not really thinking too directly on it, an answer will usually come to me. Sometimes it's a good one.

This idea came almost fully formed. I printed out a few copies of the architects' drawings of the outside walls of the house and set myself to sketching. In about an hour I had a comprehensible drawing of the longest wall of the project over which I had superimposed an enormous truss supported on each end by two piers located about four feet beyond each end of the building.

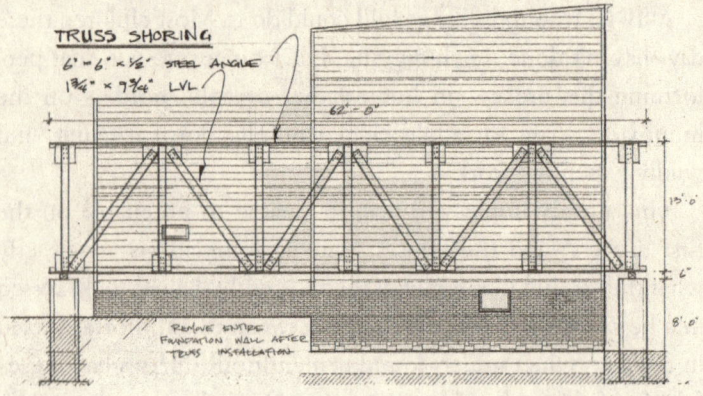

I put the drawing in a folder and brought it to the next design meeting. It took a little explaining.

"It's a truss, like in a bridge. It's strong enough to support the whole side of the building! The ends will sit on temporary piers. That way, we can demolish an entire wall of the existing foundation, dig with machines, and pour a proper monolithic foundation. Then, we take the whole thing apart and work our way around the house doing the same thing until we have a new solid foundation. We'll use it on the longest walls first and cut it down to fit the shorter walls as we go. It will give us a much sturdier result and take half the time."

Looks of incredulity spread quickly around the table. I'm used to this sort of response, so I continued, "The materials will be practically free; we can reuse the steel angles to support the perimeter walls while we replace the house's crooked old joists with the same engineered ones we'll use to make the truss. We'll send the steel back for recycling and the LVLs become part of the new building, like Mike Mulligan's steam shovel." Nobody knew what that was. "The kids' book! The one where the old steam shovel digs the foundation so fast it forgets to leave a way out. They turn it into a boiler and it lives a new life keeping the building warm. It's transformational." People who know me usually start ignoring me when I talk like this.

It took a bit more coaxing, but after a few meetings with my favorite structural engineer that resulted in a formal set of drawings complete with calculations, the architects finally came around. My engineer was tickled by the approach; most people ask him for the same old thing, over and over. He knew who Mike Mulligan was.

We are in the middle of assembling the first truss as I write this. It is enormous and impressive. It has attracted enough neighborhood attention that we are concerned that the press might start snooping around.

By we, I mean me and the other builders who are onsite every day. We are the ones who deal with the frequent visits from the Department of Buildings, the Department of Transportation, and neighbors irked that we occupy a few parking spaces (for which we pay the city's usurious fees) that they are accustomed to thinking of as theirs.

Our architect shows no such concern for maintaining a low profile. A few days ago, he posted a heroic photograph of the truss on his social media site with the caption "Nothing to see here, folks!" To drive his point home, he tagged the local neighborhood association. Somehow, he neglected to mention that the truss was a solution in which he had no authorship to a problem that he and his engineers had created. But, given the trying encounters we have had with the city and our neighbors, I'm glad my name was kept out of it. I can only guess that community relations and ethics are not required courses in Yale's architecture program.

In about a week, the truss will be even more impressive; we

will knock out a fifty-two-foot length of the existing foundation from under it, and the creaky old building will hang there until that foundation wall is poured anew. Then we will do it all over again. Twenty years ago, I couldn't have come up with the idea. I think it's majestic.

This is how I make a living. Often there is a far better, more cost-effective solution—in this case: knock the building down. The Landmarks people wouldn't allow it. We even investigated having the building condemned, but that would have required that the new structure be completely redesigned to modern codes; the cost would have been staggering, not so much because of the code requirements, but because of the architectural fees.

As long as I think of my job as a complicated puzzle, the absurdity of it rarely gets me down. I never build the same thing twice, and I never know what my next job will be. I can't remember my last boring day. I'm happy to keep at it indefinitely. There are a host of things I want to build in my workshop; some could be called vanity projects, things that have been gnawing at me for a long time. But even if I reach the point where I no longer need the money, I plan to keep doing what I do. I suspect I'm like one of those aging boxers who just doesn't know when to hang up the gloves.

Recently, I called my mother to read aloud the section that describes her and ask whether anything registered as untrue. Misrepresenting anyone can be perilous; misrepresenting one's mother is unforgivable. She only asked that I change "Paris" to "New York City" in describing Mrs. Macro's buying trips, and "cashmere" to "silk," as cashmere is not a fabric per se, and even if silk isn't exactly one either, it is her favorite. So my foray into writing has lost a smoky wisp of Parisian romance, one that I am restoring backhandedly, masked as what people who don't like to make up their own euphemisms call "full disclosure."

When I called, I still had a few more ideas I hadn't yet conveyed in this book. I realized that what had begun as a record of my time in the trades had blossomed into an examination of the qualities demanded of people who want to turn away from the life that is expected of them and build the life they wish to live. I had been doing a lot of thinking about what that takes, and of all the people who have inspired me on my way, my mother continues to be my primary example of someone who lives life on their own terms. Industry, resolve, fearlessness, indifference to approval, self-reliance, optimism, and even cussedness combine in her to make what I believe to be Will. In all our years of conversations the topic had never come up. I know I am given to flights of fancy and sought her more measured opinion, so I asked her, "Mom, do you think it's fair to say that people who have developed Will can see the future?"

"Oh yes, absolutely, dear."

She calls me "dear."

My mother is one of the most pragmatic people I have ever met. She is the daughter of an agricultural economist and a statistician; frothy spirituality is not part of her genetic makeup. Based on her confirmation, I feel confident in saying: People who have developed Will can see the future.

The process is simpler than one might imagine. The popular belief is that developing Will is a matter of visualization. That is a gross misrepresentation promoted by people who wish there were an easier way, or are trying to sell great gobs of toothpaste. You can visualize your ideal life until the day you die; the practice won't bring anything into being. Of course, it's better not to figure this out on the last day.

I'm a firm believer in visions; well-formed ones are rarer than you might imagine. But every vision comes with a price tag. If the willingness to pay the price in tireless effort that realizing a vision requires is not passed genetically, at least the impetus for it can be passed by example. It is first among the gifts my mother gave me. The second was just enough self-respect and stubbornness to turn away completely from the prudent path she had planned for me.

Like everyone else in the world, I was born with the ability to do almost nothing. I was fortunate to be raised among people who were unafraid to hazard the improbable. Effort is its own reward, an effort made one day builds the strength to make another more easily and more effectively the next, and after a few thousand days of effort upon effort, things will have changed, completely.

I hope that in writing this, I might help someone find inspiration and perhaps meandering guidance from the travails recounted herein that are happily mine. The stories were conceived, remembered, and written one by one as bedtime stories I might read aloud to a friend. I have arranged them into chapters that are titled by the central concepts that have been meaningful to me on my way. The experiences, efforts, and people they describe make up the saw-toothed arc of development that my life has followed. To me they are parables, each containing a few simple lessons. They are presented as such in the hope that they may sound sympathetic notes in a reader or two. Each of us faces unique difficulties, gets tripped up by our shortcomings, stumbles and blunders through life's challenges. With luck and determination, it is possible to make our own way. This is the story of how I made mine. Inspiration and guidance are all I can offer; no one can make any effort but their own.

The future is, for the most part, as invisible to me as it is to everyone, but I have sown a few seeds of things that I mean to bring to fruition. These are the things I can see clearly: I see the completed homes of my current clients in almost every detail. I see personal projects on my workbench that I make only for the satisfaction of doing so. I see the granite-faced firehouse I just purchased with a shiny new cast-iron storefront and tasteful silver leaf sign. There are a lot of things about the future I see but won't tell; they are *my* business.

Work hard. Do what you want. Don't dare flatter a child; they may never recover.

HOW
TO
BUILD
IMPOSSIBLE
THINGS

CHAPTER 1

Belief

I believe I am things that I am not, and have disbelieved
I could be things that I assuredly am.

Like all serial dropouts, I spent the first few years of my career flailing about. I lived in a string of cockroachy apartments, took work where I found it, and slept on enough couches to develop a liking for it. An artistic bent mesmerized me into thinking that was the way I was ultimately headed. My guitar accompanied me everywhere; I painted pictures, carved stones, and even scribbled down a few listenable songs.

A neighbor on the Lower East Side belonged to an itinerant theatrical troupe. They could be found every weekend in Washington Square Park, encircled by gawkers, tossing around old vaudeville jokes and juggling clubs. He would spend evenings teaching me circus skills, encouraging what he took to be my "naturally performant" side.

Despite my misgivings, my neighbor decided I needed to get out there and take my place on life's stage. My only experience in front of audiences until then had been in sweaty-palmed trepidation of my turn at the piano recital bench. I had no memory of having basked euphorically in any audience's adulation. A knotted stomach and wish to disappear was all I could recall. But my neighbor maintained that nothing matched the energy exchanged between audience and performer; in his estimation, it verged on the sacred, or at least the narcotic. He had a practiced exuberance and an endearing tenacity. After several evenings of caviling, I was convinced enough to play along.

December of that year came with its damp New York cold. My neighbor knocked on my door and announced that he had a gig for me. An upscale Fifth Avenue department store was running a promotion for a denim-centric clothing line with a western theme and they needed a banjo player to set a suitably oaty mood. Nothing about this announced itself as my big break. I had hiked Fifth Avenue's canyon and felt out of place amid the Saks and Pecks and Taylors. Moreover, I was really more of a banjo owner than a banjo player. My entire oeuvre consisted of a decelerated version of "Foggy Mountain Breakdown" and the opening theme from *The Beverly Hillbillies*. Fashion had eluded me all my life, and even more markedly in my seventies adolescence, when the only flair I exhibited was for owning the wrong brand of everything.

My neighbor lured that they were paying $250 for two hours' work. I took the bait.

The appointed day arrived; mushy, gray, and dank, it was well matched to my expectations. From the meager offerings of my closet, I cobbled together a poor man's cowpoke ensemble: boots, bandana, jeans, and a piped shirt. My banjo, freshly strung and inadequately practiced, was secured in its case and tucked under my arm for the trek uptown, the handle lost long ago.

I had been directed to appear at the store's service entrance, my first in an entire career spent going in the back way. I was bounced from Security, to Personnel, to Sportswear, and finally found a familiar face: a clown I knew from the theater troupe who doubled as their booking agent. She took my arm and led me to the "artistic coordinator" who was to instruct me in my duties. My clown friend wished me luck and passed me the cowboy hat she was wearing, saying, "Looks like you need one of these." Clowns are often awkwardly kind. I tried to twist it down around my brow; it was ludicrously small. I perched it on the back of my head in the manner of unthreatening buckaroos and felt the gnawing creep of coming discomfort.

The artistic coordinator led us to a dressing area. A piano player of considerable skill was wending his way through a catalog of singing cowboy classics. "You know this one?" he'd ask, before launching into a swinging-doors version of some lost ode to manliness and verve. "Mmm," I'd answer, "that's a good one." Nearby, three models were busily assessing their self-worth in banks of mirrors set up for the purpose; the first wore a tailored tennis dress, the second a feminized sweat suit, and the third an ensemble of skin-tight, eighties-embarrassing workout togs. Something was amiss.

The artistic coordinator gathered us up and announced that the theme of the promotion had changed: "Jesse's is super excited to introduce their new sportswear line, just in time for Christmas! Ted, you play your piano here by the sales desk; Christie and Melanie will stay by you. Jill, you walk the floor; Mark will trail you with his banjo attracting everyone's attention!"

What a plan.

To her credit, a shapely eighteen-year-old in neon spandex trailed by a red-faced cowboy with a two-song repertoire and muddy boots did elicit quite a few stares. But I doubt that all of our agonizing efforts, stretched through two interminable hours, produced a single sale. By the time it was over, no one craved an early retirement like I did. I hustled back to the dressing room, packed my banjo, found my clown, collected my check, and turned in my cowboy hat.

It is a peculiarity of my psyche that discouragement found on a given path rarely leads me to abandon it entirely. My history has been one of probing about for possibilities. Perhaps if performing wasn't exactly my cup of tea, something more suitable could be found. Thus began a lengthy list of investigative career choices, some taken on out of interest, all driven by the incessant need to eat. Many ended with a dramatic flourish:

- Ice Cream Server and Cake Decorator. Madison Avenue and Seventy-second Street. All went well for a year. I had the distinction of decorating several cakes that were picked up by Patricia Nixon as special treats for her husband, who was in regular need of sugary cheer. When the store's lease was up, the rent was raised to untenable. Ice cream was out; Ralph Lauren was in.

- Bindery Worker. Cedar Street. This firm printed analog stock reports for Wall Street and knew better than any I've encountered that a deadline unmet was a reputation lost. All-night mailings were a weekly ritual. I lasted as long as I could, but after the alcoholic breakdown of a machine operator that involved a crescent wrench and an automatic envelope stuffer, the cleanup of two severed arms when a sleepy laborer was sweeping the table of a hydraulic paper cutter whose safety switches had been taped down to speed work, and a hallucinogenic all-night letter-stuffing session fueled by sugar-cube-sized blocks of hashish distributed by the boss's son, I was through.
- Animal Food Delivery Driver. Twenty-sixth Street and Second Avenue. I was at the age when the weekly appearance of Irene Cara in the store's aisles was an unparalleled thrill. I can't remember seeing anyone like her hanging around the steel mills back in Pittsburgh. Things fell apart when one of the managers started a cocaine delivery service that ran in parallel with the animal food offerings. After two police stops for my driving irregularities, I lost my nerve.

Finally, I found work as the assistant to a nomadic Armenian American carpenter who spent his evenings as an unapologetic sexual sadist. Why anyone would find carrying a hundred 70-pound sheets of drywall up several flights of stairs on a New York summer's day appealing is anyone's guess, but I had found my place. Work that most people would call brutal I call satisfying. I still enjoy the fatigue and deep well-earned sleep that only a difficult day's labor brings.

So I kept at it, from company to company, and project to

project. Every job has something to teach, even if it's "I never want to do *that* again." But most of them have added a useful arrow to experience's quiver; I still find the occasion to use a cake-decorating technique when applying caulk in oddly shaped spots, and I still stuff envelopes in the peculiar way we did it at the bindery. I've even been known to grace a stage from time to time when a country-style musical performance is required of me, although I still don't get the audience energy thing, and I refuse to wear a cowboy hat.

"Belief" and "sacred" are two words that appear together a lot. One would think that people spend long hours pondering life's puzzles, carefully weighing arguments and rebuttals, measuring each against their experience, then rigorously holding both to the clear light of reason. This presumption differs markedly from the unfocused capers of my internal mullings. My pronouncements on life are often haphazard, drawn from sources ranging from kindergarten companions to country music legends. I can say with confidence that most of them have benefited from less pondering than tonight's choice of dinner.

From time to time, I've been jolted by circumstance and chance into admitting the falsity of what I call "my beliefs." Life's tsunamic changes can bring about unpredictable behaviors in a person. I am not privy to what takes place in most other people's hidden lives, but in my own, a thorough dunking and cleansing in the worst of life's currents, no matter the discomfort of it, nor the dismay it brings, washes clean whole attics of nonsense, and makes some seldom-visited rooms suddenly inviting and inhabitable. Nothing brings a fresh perspective like good, harrowing upheaval, although I can't recommend it as a regular practice; the pain of it mitigates my appreciation of the eventual result. There is no getting around the fact that in every life, terrible things will happen. Defeat, loss, pain, and death are our shared burden. I doubt their scars ever disappear any more than the scars I wear from my physical injuries have. And I doubt I will ever welcome them—I don't think I even want to—but I have endeavored to look at, listen to, and examine the most difficult episodes of my life. Without fail they have shown me that what I thought I believed was inadequate, cowardly, and confining. I am not the tender soul I was once content to believe I am.

Mercifully, I have found gentler ways of investigating the contents of my character. There are three or four people in my life whom I respect above all others. Each of them is in the habit of pointing out the more suspect of my opinions and pontifications. Because I care for them, and know they care for me, I try to make enough room among my crowding thoughts to allow that they might be right. After a few days' consideration, it more often than not turns out that they are, in fact, correct; I have been a fool again. Another piece of fatherly advice or insight borrowed from an old magazine gets consigned to the dustbin.

By firmly believing this position or that, we wind up not believing enough about ourselves; we exclude everything we cannot bear to be true. It's not just the embarrassing, painful, and shameful that we won't admit; there is a good deal we exclude that is pleasant, even joyful. It may be comforting to say, "This little road map over here is a diagram of the good and just person called *me*. These are the paths I always take, and these are the things I hold dear." May I suggest that there is an awful lot going on both internally and externally that this approach will never satisfy.

I offer a short list of things I no longer believe:

I am a nice man.
I am honest.
Love lasts forever.
Life goes on.
I am not an artist.
Everyone gets to make their own list.

I may not be nice, but it pays to not be awful. Sometimes this goes against my natural inclination. This is the story of how Maya saved me from the Buddhists. I offer it as partial repayment on a debt of kindness. The kindness in the story was not mine.

Manners are my first defense against my rougher impulses, but they are easily overrun. I think and talk badly about people once they've left the room; I heap derision on their ideas, sometimes out loud, sometimes just in my head, where derision for heaping is easy to come by. But people have more redeeming qualities than I allow, and it pays to treat them with deference and respect, despite whatever beliefs I might hold.

About twenty years ago, I was given the chance to do a Park Avenue apartment renovation with my first "name" architect. Since I'm the type to rarely recognize an opportunity when it presents itself, I didn't give it much thought. The plans indicated a "minimalist" design, meaning the draftsperson had figured out long before that although computers struggle with irregular shapes they draw rectangles like nobody's business, so a colorless 3D Mondrian is what the apartment was destined to be.

Minimalist spaces are among the most difficult to build. It seems counterintuitive until you actually build one. The walls of this apartment were half defined by banks of cabinets, closets,

and appliance coverings, all veneered with six-inch strips of elm
cut from a single tree. Swinging, sliding, and hideaway doors
were covered in the same material so that whole quadrants of
the place looked like a seamless repetition of the same wooden
pattern, each panel having a different function or leading to a
different room. The pattern was uninterrupted by visible hardware or distracting wall switches. A first-time guest would find
it difficult to find the fridge or get to the bathroom in a hurry.
Various types of flooring—walnut, granite, and concrete—were
laid to delineate the different spaces, perhaps as a visual cue
along the lines of "Oh, concrete, I must be in the kitchen!" The
floors met the walls and corners with the same regimental regularity. Joints between slabs of stone or floorboards were placed to
precisely align with the seams between the elm veneers. Like the
foot of Cinderella's taller stepsister, everything was shoehorned
into this schematic slipper, or unceremoniously lopped off and
forced to fit.

Drawing such a place is simple: Line up the corners of all
your rectangles and, voilà, the composition is complete. Execut-
ing the construction is another matter altogether. The only way
the illusion works is for each trade to place every element in its
precise location front to back, left to right, and up and down.
That means that at every conjunction of materials—for instance,
where woodwork meets a window—every component must be
planned, manufactured, and installed to within a few hundredths
of an inch. The window, its surrounding frame, the stone sill,
and the woodwork are all made in different shops, none of
whom communicate with one another. Those shops will send
four teams of installers, who will arrive with no understanding
of the architect's rigid design intent. All of this takes place in a
prewar building designed with no such level of exactitude in
mind, and in an apartment that has usually suffered the indigni-

ties of several renovations. A misalignment of a few millimeters in any direction and the spell is broken. It doesn't take an artist's eye to notice when three or four lines don't converge exactly. Getting the thirty fabricators an average job requires, with wildly varying levels of skill and interest, to adhere to these standards is a maddening task.

Behind all these unforgiving walls was a snarling maze of wires, ducts, light housings, transformers, waste and water pipes, smart home cables, scenic lighting controls, shade motors, projector televisions, invisible hardware, pivoting cabinet assemblies, and all the other trappings that make a modern Minimal home.

The grunt work of producing drawings for this place was done by a young architect of working-class Brooklyn beginnings. He was in a bit over his head and was happy to let me pitch in on the myriad details that needed sketching. He behaved in many ways like us humble carpenters, so we got on. He worked for another architect who would come around from time to time to check on our progress. This gentleman displayed more of his vocation's affectations: His language skewed scholastically repetitive in the way Ivy Leaguers hope might be just outside their listener's comprehension level, his demeanor rode the line between deference and dismissiveness, and he wore architect shoes. Despite my uncharitable first impressions, I came to admire him in time.

That architect worked for the visionary architect behind the project. She had long been a luminary in her profession and consequently provided the "name" with which we opened this story. Our visionary was slight, nearly always clad in blacks that matched her shiny hair, and had a way of walking that approached frolicking. She never carried any papers or made any drawings, but rather swept her arms to indicate that a constellation of lights was to go here, or lowered a knifelike hand to show that a keypad was to go there. She had an energetic confidence that she *would* make her vision come to life, even with louts such as us, so I will call her Maya. She was always friendly, and delightful in her way, but we were mean and men, and would make fun of her as soon as she disappeared into the elevator.

Maya was right; that job got built.

Five years passed, and five more renovations came and went. I took my first and only office job, working as a project manager for a general contractor. From my chair, I learned the ins and outs of the financial side of construction and gained weight. I was assigned to the total makeover of a penthouse apartment in one of Central Park West's idiosyncratic old towers. It belonged to a real estate mogul and his wife who were rounding the last bend of middle age. Their children had fled the nest, and this was to be their sunset aerie. They were prominent Buddhists; they had even met the Dalai Lama, who is apparently more worldly than I had taken him for—just causes don't prosecute themselves for free. We didn't see much of them during construction, only at monthly requisition time when the husband would walk through the project with me and, in the most measured of tones, contest the degree of completion of every item he found on the bill.

Minimalism had tightened its stranglehold on the high-end world; years had passed since I'd seen a newly installed wooden molding. But every building style has its challenges; that's where the fun lies, so I wasn't exactly unhappy. We installed sliding computer-carved op-art shoji panels, luminescent resin sinks, mesh nickel fire screens, colored concrete counters, and a host of other contrivances for readers to drool over when their future issues of *AD* arrived. I suppose *Architectural Digest* had become too unnecessarily cluttered to spell out in full. Minimalism had taken root in the publishing world as well.

Things went acceptably well until move-in time. Having a fancy New York apartment built really isn't much fun. It's dirty, noisy, takes too long, is inconceivably expensive, and rarely finishes on time. Moving in always brings a fresh cartload of unpleasantness. Owners can't be blamed for nearing their wits' end. Carpenters were still lunking about, installing final bits of hardware. Things were being polished, removed, replaced, and corrected all around the place. Who would want to bring their prized raku collection or custom handwoven Tibetan carpets into such an environment? We did our best to keep things clean, and to hurry ourselves out the door, but it did little to stop the lady of the house from screaming herself hoarse every day.

Unfocused screaming is difficult to address. Had there been specific needs we could have met, or tasks we could have performed, things might have eased, but in this instance, *everything* was wrong. The apartment was poorly built; our standards were nonexistent; we lacked manners and any sense of appropriate behavior. Her husband's rants were more specific: The through-the-wall air conditioner made a distinct hum that made it impossible for him to use his Meditation Room, and we had stolen a set of handmade Danish silver.

Buddhism was not doing any of us any good.

The tension built; I dreaded every day in the place. The only upside was that the ceaseless stress ruined my digestion; I finally lost some weight.

One afternoon, I was informed that we were to clear out early; a friend of the owners' was coming by, and we were to be sorted and gone by three o'clock. My crew left at two-fifty. I was putting on my coat and about to make for the service entrance when the main elevator of the apartment opened. I was standing ten feet away; the wife was between me and the elevator door. I looked to escape when I heard a warm cry from the elevator: "Maaarrrrrk!"

To the wife's horror, Maya appeared, sleek as ever in her jet-black jacket. Maya breezed right past the owner without a glance, threw her arms around my neck, and cheerfully greeted me with "How are you?," "It's been so long!," "Did you build this place?," "Oh my god, it's soooooo beautiful!," and to my client: "You are the luckiest person in the world to have him!" She took me by the arm and made me show her every scrumptious inch of the place.

I didn't deserve it. I had been pleasant and polite with her and chatted amiably if she wanted when we had worked together, but I had said and thought enough meanness to outweigh it badly. And here she was, saving me more completely and earnestly than I ever could have hoped for. I can never repay her.

We didn't hear another cross word out of that pair; a strange serenity came over them. They would even let us stay when friends came for a tour. Maya's seal of approval had transformed their new home into everything they had hoped for. Maya transformed me as well; I don't even mind having to taste again the shame I feel at mistreating her, whether she noticed it or not.

A few days later, the silverware turned up, too.

An untested belief isn't worth much. It's tragicomic that I often hang my existential hat on bumper sticker slogans, well-meaning advice from advisers who barely knew me, and overwrought song lyrics from the seventies.

We are all confounded by:

What we believe life is.
What we believe we are.
What we believe we are capable of,
And what we believe we are allowed to do.

We work so hard to pin ourselves down. There's no shame in letting some of it go.

+ + + +

CHAPTER 2

Talent

When inspiration strikes, do that thing.

I have a bone to pick. Talent is a subject that has dogged me my entire life. As a child, and even as an adult, whenever I displayed an ability that was any measure beyond banal, I was met with exclamations that invariably contained the words "he," "is," "so," and "talented."

"He"—In my day, there was little room for questioning this in a public way.

"Is"—Seriously? I'm supposed to explain existence in a single sentence? . . . Live fully or hide.

"So"—In a world where the extraordinary actually exists, a finger painting bearing the slightest resemblance to an elephant would benefit from more precisely meted praise.

"Talented"—Are we to understand that a creator sows seeds of destiny in each of us that, as we grow, will blossom into the symphonic maestro one of us is meant to be and the fry cookery that is another's calling?

Even if it is true, I hope it's not.

In my view, discussions of talent suffer from two simplistic misconceptions:

1. Talent is the most significant determining factor in predicting achievement.

    Inborn ability is a difficult thing to measure. IQ tests are an effort to do so. Correlations have been found between income and IQ, but such studies neglect all the washed-out geniuses out there who never found their way to success. It's the old correlation/causation problem high school debaters often get mixed up about.

    Other areas of supposed innate talent aren't measured at all. How much latent ability does a musician or a novelist have before they begin their studies?

    I'm not entirely discounting the inequity of the genetic roulette spin, just positing that from there one takes one's winnings to the developmental craps table.

    Of the two children my age who were the most praised for their talents, one became the head of an innovative global medical research institute, and the other died of an overdose beneath a highway overpass.

    There's a show on television these days called *America's Got Talent*. I've only seen it once or twice, but I really wish it were called *America's Got Quite a Few People Who Have Been Diligently Practicing a Single Skill with Admirable Regularity*.

    Now, that's a show I'd watch!

2. If one is not immediately good at something, it's not worth doing.

    Thoughts don't get much dumber than this. Interesting people are interesting because they have interests—interests they often pursue simply because they find them interesting. If our method of education weren't so busy ranking and comparing children all the time, kids might find it a bit less uncomfortable to bear the mediocrity they so often exhibit. Is it really so insulting that I am the eight hundred thousandth best guitar player in the United States today? I play every day anyway. I'm not expecting a trophy.

People bristle when I talk about this topic, so I will try to say it all over again in a different way. I repeat myself because ideas about talent and ability are stumbling blocks my acquaintances speak about almost every day. People feel bad needlessly about talents; they think they have none or have fallen short in their development. It really doesn't matter how good a person is at something; what matters is the meaning they derive from its pursuit.

"Talent" is the great sifter. We use it to sift ourselves, to decide what is worth pursuing and what we should abandon after a few tries. Never mind that most of the truly accomplished individuals will tell you, even on TV, "I had no interest in vaccine development as a child" or "It was eight or nine years of rigorous training before I could put the shot worth a damn." No one listens to these people. We listen to second-grade teachers who praise the sparkling talent of some little darling who arpeggiates their way through "Für Elise," or trills the life out of "America the Beautiful," and we decide we will never venture another note.

At nine or ten I heard the story of Wolfgang Mozart writing concerti at an impossibly young age. There are handfuls of examples like him scattered through history, just as there are handfuls of billion-dollar lottery jackpot winners. Statistics teases most seductively with its outliers. Little was said of papa Leopold, but it certainly helped Wolfgang along his way that his father was a composer, performer, and teacher who had written the definitive text of his time on violin technique. Leopold was also a taskmaster who required hours of daily practice, beginning as soon as Wolfgang could reach the keyboard. I mean to take nothing away from Mozart: He pursued his passion tirelessly, even to his own ruin, and brought previously unimagined possibilities to music, freeing it harmonically, rhythmically, and melodically, imbuing it with an emotional breadth that precious few have managed.

The halls of history are littered with people of wondrous talent who never became anything or found wreckage in unendurable fame. Talent, no matter how spectacular, is often more liability than asset unless it avails itself of benevolent parenting, rigorous practice, wise instruction, and the near-impossible feat of levelheaded humility. For all of life's "natural talents," there are countless examples of individuals who practiced harder, explored every avenue, and became great by work, focus, and grit. To attribute Mozart's accomplishment to "talent" is insulting to everyone's possibilities; it leaves no room for them to follow in his footsteps. What would be the point? Only he was born with those special shoes.

Every childhood has its kernels of prospect. I must have been four when I would lie under the piano in the evening while my mother played, bathing in the tactile vibrations that marvelous wooden and metal machine could produce. Almost every fascination I carry to this day can be found in that scene. Tens of thousands of hours of practice, scores of teachers, numberless failures, and a handful of well-completed enterprises separate me from and unite me with that boy. It doesn't matter what degree was reached in the pursuit of those fascinations, or the degree of innate ability with which I may have been blessed.

Circumstance, physiology, and temperament aren't ours to decide. Pursuit and practice are. The joy of following something as far as one can is enough. Every endeavor is endless. No one ever gets to the bottom of anything.

Did I have "gifts"? Did I have "talents"? Certainly. Does every child? I hope so. But the surefire way to extinguish them is to tell a child which "talents" they have and which ones they don't. I hate the word. More joy and real self-satisfaction never happened because of that word than any other in the English language.

My career in craft began in my father's workshop. I'm not sure I was allowed to be there alone, but tools and wood have always drawn me. I was not a prodigy, but early on, I exhibited curiosity and stubborn determination that have both dogged me and served me well.

When I was five or six, I found myself in the basement of our house in Pittsburgh. There was nothing particular to do, as thankfully, there usually wasn't in those days. I pulled out a few of my father's tools, rummaged through his wood scrap bin, and set to work. I have no sense of how much time elapsed, only that I smashed my thumb with his hammer at least once and cut my hand on the edge of his coping saw. None of that is important. Healing, although not entirely reliable, is the greatest benevolence this world has to offer. The result of my efforts was a poetry-book-sized platform with four stubby legs, in the center of which there was affixed a wooden crank made of dowel sections, a single nail, and a short pine arm. It worked. For real! It cranked. I could grab it by the handle part and spin it around, cranking here and cranking there. It could crank anything! I went through the house, cranking everything that needed to be cranked, threw it in my toy chest, and hardly gave it another thought. I'd come across it from time to time while on the hunt for something else. I suspect that when we moved from Pittsburgh, some parent, in their benign neglect, tossed it out for good rather than pack it off and pay for its shipping. That's fine . . . to be expected. Children of my generation lacked the magicality of today's offspring, who somehow elevate everything they touch to excruciating preciousness.

As much as everything we have is lost at last, nothing can ever really be taken from us.

MIDNIGHT AT LUIGI'S

Beginner's luck is a real thing. Perhaps it's the universe's way of providing some encouragement to those who are willing to try new things. Perhaps it's a way of taking the smugly established down a few pegs so they'll renew their efforts. Perhaps it's that we celebrate and record the rare occasions when we get something right on the first try. I don't know. I don't make the rules.

Twenty years into my career, I had at least touched on most aspects described by the umbrella term "carpentry." A new millennium had arrived; we had all escaped the latest predicted Armageddon, this time advertised as the inevitable result of a worldwide computer glitch. I turned thirty-eight, just as I had been calculating ever since I could remember knowing that a year with three zeros was coming. A whole new era was here, and nothing terrible had happened yet. The world felt hopeful.

The contractor I worked for then was known for taking projects with complex, innovative elements. In practice, this meant that many aspects of the jobs were working prototypes that the owners would live with, use, and hopefully enjoy. The single oddest aspect of high-end contracting is that our clients demand innovation and uniqueness. The cutting-edge, "only-I-have-one-of-these" elements of their homes are the ones they show off first to anyone worthy of a tour. Very few architects remain who are capable of engineering and designing singular structures like this, so it falls on the contractors to produce unique elements that are often no more thoroughly conceived than a line drawing depicting their intended shape. I have come to see this design gap as my personal niche, although it is obviously a precarious career path. I just happen to like building things no one has built before, and I have for as long as I can remember.

For two years before being hired by this contractor, I had worked for one of his subcontractors installing architectural woodwork. I had developed several unusual systems that sped my work along without interfering with my pursuit of quality. This boss took notice, and eventually pulled me aside and openly wondered what it would take to make me jump ship. Several hard swallows later, I was hired and sworn to secrecy.

My contractor boss liked to call me his "technical specialist." We settled on this title after I assured him that calling me an "engineer" was patently fraudulent. Engineers have degrees, pass exams, and are issued official-looking stamps that hint at some level of familiarity with the laws of physics. A GED issued by the great state of Wisconsin bears no such authority. I was assigned to come up with a scheme for building all manner of novel assemblies: an ultrathin floor structure for an added mezzanine in a Tribeca loft; a monumental sculptural six-story staircase in a Park Avenue townhouse; a sweeping Serra-steel fireplace spanning the living room wall in an Upper East Side townhouse; a Transformer-like office nook that changed from a simple cubic room into a workspace with a rotating plasma TV, a drop-down worktable, and a tricked-out computer station.

I would hand draw, in ink, the components of each of these

assemblies, source the hardware and materials, and send everything out to the varied fabricators who helped make them.

Drawing is how I think about building, and often how I talk about it. Drawing in ink is how I make sure I mean what I say. For the first twenty-five years of my career, I worked with architects who pondered and communicated in this way. We spent long hours working through a project's most complex problems, sitting together around a plywood desk. Those architects carefully drew each layer, every screw and bolt, erasing and reworking a drawing until the problem's solution was clear to both of us. I adopted the habit, often copying their style to give my work some flair. In time, while my hand drew objects in two dimensions on paper, my brain began to assemble them in three dimensions in my head. After several years of practice, I became confident that I could figure out how to build even the most inadvisable of ambitious architectural assemblies. Whenever something felt outright dangerous, I would send my drawings to a freelance engineer, Lenny, who would look them over and stamp them for a few hundred dollars. With Lenny's stamp of approval, we could all proceed in the overconfident belief that "due diligence" had been met.

Most things worked. Sometimes a door would warp and need straightening, so I invented a method of invisibly installing a guitar-style truss rod to pull it straight. Once or twice a mechanism wasn't quite as robust as it should have been; I'd have to go back and do a surgical repair to make the thing work reliably. No fix has ever taken more than a few days, so for someone regularly sending prototypes out into the world my success rate gives me some assurance that I am on the right track. But the fear of catastrophic failure has been a constant for me; even with our engineer's confirmations, it has kept me awake many a night with visions of fires burning out of control, or an extended family assembling on their staircase for a Christmas photograph when a button-nose niece proposes that it would be fun if they all jumped at once.

Some of my job's challenges were technical; some were artificially introduced when unlikely deadlines were imposed. Some jobs had both. No matter, I loved the challenge, and I was just enough out of the line of fire when things went awry to keep the stress bearable.

A few years into this routine, we were awarded a project in a newly built Chelsea luxury development. The top two floors were ours to finish. They belonged to the developer's son and his future husband. The developer was a man in his seventies who had arrived from Italy as a young plasterer and worked his way right past every boss he ever had. He had become a little hunched and puckered, but he was strong and forthright, and showed an easy love for his son that I found touching. His son Luigi was strapping, darkly handsome, with an open gruff voice that allowed conviviality but warned off misbehavior. I like liking my clients, even better if I respect them.

The designers of the apartment had conceived a blending of traditional and forward-looking elements. Some rooms were small and intimate, like the silk-and-oak-walled reading room that was approached through a complex hidden door. Some were expansive and opulent, like the grand living room with its oversized plaster crown and antique mirror-backed bar with Giacometti-like columnettes separating bronze-rimmed leather panels. My initial skepticism at the approach was unfounded. The ordered elements grounded the more fanciful ones; the married opposites embodied in the apartment's design were neatly suited to the starkly disparate personalities of its owners: solid, serious Luigi and flighty, dreamy Sean.

Working on their home was instructive, not just because it

broadened my staid ideas about design, but also because the designers and clients were decisive and demanding, characteristics rarely found in tandem. They wanted the place finished quickly and they were not going to be the ones standing in the way.

It is not possible to custom build several million dollars' worth of house without something going wrong. This one had its share of missteps. A plumber cut a two-inch gash clear through a solid sterling silver sink. My project manager hired a silversmith friend to repair it. He tucked himself into the kitchen cabinet beneath the sink and began soldering. From inside the cabinet, he was blissfully unaware that his torch had ignited the red rosin paper protecting the walnut counters. I will ever remember with amusement and horror the image of our gangly laborer Stitchy dancing a limberjack's jig, extinguishing a leaping flame with each stomp after he swept the paper to the tiled floor. Happily, his gyrations wound down successfully before the sprinkler system went off.

As the deadline neared, everyone was aware that we were cutting it close. Armies of movers, picture hangers, carpet installers, drapers, cleaners, stagers, and decorating assistants had been scheduled long in advance, and no one was entertaining changing their arrival date. Luigi was a builder's son and had some idea what this meant: more workers, more hours, more days, more pressure. We called in reinforcements from other projects. Luigi arranged for us to use a section of the building's underground garage as a temporary workshop, with no restriction on hours or elevator access. The end was coming fast. We bade farewell to our families and encamped.

Pizza and egg sandwiches became our staples. My contractor boss paid for every meal, knowing it would boost morale. I assigned tasks on a minute level: straighten those hinge screws, polish that scratch, nail up these three pieces of molding. Twenty-five carpenters, five laborers, and a pizza-purchasing boss can get hundreds of things done in a sixteen-hour day. I ordered all the missing pieces from wherever we could get them; things really were shaping up.

The apartment had a simple internal staircase leading from the foyer to the upper floor. It had already been installed, but it lacked a handrail. The designers had specified a classical profile from a company that made that sort of thing. We needed a spi-raled volute to initiate the run as a complement to the curved bottom stair, a few straight sections, and a turn where it bent to meet the wall at the top. I am not a qualified metalworker, so I consulted a friend who apprenticed in French blacksmithing to make sure I was buying a weldable alloy and I asked if he would join the pieces together for us when the time came. He was more than happy to help; I'd brought him on for some bigger projects in the past that were more suited to his impressive skill set.

Two days before the movers were scheduled, the pieces finally arrived. I set them out along the stairs to make sure they would work and realized immediately that I didn't have a piece that would transition from the level volute to the raked rail that ran up the stairs. My heart sank.

Deadlines have no patience for heartache, so I gathered my-self up and took a hard look at what needed to be done. I needed a $\frac{3}{4}$" × 6" × 12" piece of brass just big enough to match the profile and allow for the gentle curve the railing required, I needed some means of bending it, and I needed a way of carv-ing the piece so that it looked like the fancywork of the railing sections I had. I set out for the street, grabbing a cab to a surplus metal store near Canal Street. They had a brass offcut of the cor-rect alloy that would work. I grabbed another cab back uptown to a well-stocked hardware store. They sold me a Dremel tool with a host of hobbyist's attachments used for shaping parts. Unconvinced that this was up to the job, I bought every burr, rasp, file, jigsaw blade, and sander in the place that looked like it might help. I left just as they were locking their door for the night.

The apartment was becoming far too finished for this kind of work, so I took every tool I might need to our workshop in the garage. It was dark, cavernous, and cold down there. A few work lights brightened the place up. Sawhorses and a piece of plywood made a workbench. I was in business.

I traced the gentle curve onto the brass slab and set about the slow process of cutting it out with my trusted aging jigsaw. Several blades later, I had my blank. Now for the bending. Unless you've worked with brass, which up till then I had hardly done, you wouldn't know that a $\frac{3}{4}$"-thick piece is all but impossible to bend by hand. I hammered it; I stomped on it; I heated it with a propane torch until it burned through my work gloves. It was unyielding. Finally, I decided that what was needed was an anvil. Parking garages rarely have anvils just lying around, but they did have a large steel dumpster, which is the next best thing. I clamped the end of my blank into the crook of its hefty handle and set about beating it mercilessly with a wrecking bar. Dozens of blows would produce the slightest hint of a bend in the brass; I was encouraged. Each time I produced a bit of an arc, I would move the piece a hair farther into the handle and resume my pounding.

Midnight was approaching. The noise a worker makes is never as annoying to them as it is to those around them. Given the level of racket I kicked up in that garage, this truism extended to every occupant of the building. I was oblivious to the vexation I was causing until I heard a voice from behind me.

"It's midnight, can you keep it down a little?" Luigi was standing in the slanting light with a half smile, half scowl. "Oh, sorry, I was trying to get the railing finished." Luigi looked at my work, the dumpster, the brass, the crowbar, and sweat-drenched me and said, "Go ahead. Get it done."

No one interrupted me for the rest of the night. The curve formed gently around the dumpster handle. I worked straight through till morning carving the fancy profile into my piece with the Dremel tool. I polished it with a succession of sandpapers until my crew arrived around eight o'clock.

My blacksmith friend showed up a little later with his welder. I had cut my shiny new piece to fit the staircase and laid all the pieces along the stairs to assure myself I had everything right this time. I did. I was proud. I picked up the little curved piece I had just finished; it was all of ten inches long and had nothing about it to distinguish itself from the other pieces. I held it out to my friend. "I made this one last night."

"You made this? Nice." This was all the praise I ever needed. He welded the sections together, polished out the joints, collected his tools, and went away satisfied.

The next day, the troops arrived. There must have been fifty of them. Dozens of carefully labeled boxes were opened, their contents neatly stowed. Drapes, art, furniture, rugs, dishes, silverware, books, all brand-new and perfectly coordinated, arrived every hour for the next two days. I had never seen anything like it. When they finished, everything anyone could need to lead a stylish life was neatly tucked in its place, right down to napkin rings and lavender dish soap. The place was finished, and it was fabulous.

On a tour of the apartment, the owners would probably show off the custom sterling silver hardware, or the nearly invisible and devilishly clever door that leads into their reading room. The designers would probably point out the dramatically lit, sculpted bronze, leather, and antique glass bar. Any tour I gave would begin and end with that one short, bent section of hand-made railing, completely indistinguishable from its neighbors.

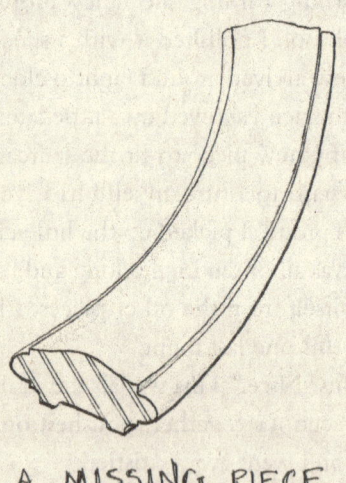

A MISSING PIECE

+ + + +

recover from the effort. It was worth it; nothing is sweeter than
a well-deserved break.

At age five, it was decided that I would begin taking piano lessons from a woman who taught privately at my elementary
school. It wasn't that I showed any interest or aptitude for the
instrument; rather, it was the fate of every child in my family to
take up piano at five. This sort of summary decision is frowned
on these days; parents seem to think that the child in question
should be consulted, or at least coaxed into any regimen. This is
hokum. Had I been consulted regarding my preferences at five,
I would have announced that it should please me to live in a
house made entirely of ice cream, and that I must be allowed to
go swimming whenever I wished. Since there has never been a
morbidly obese swimming champion, this path probably would
not have been a star-strewn one. So piano it was.

Teachers should not take their responsibilities lightly; young
minds are at stake. Steady, self-assured guidance can bring wonderful results. At the same time, expectations can be firmly set
without instilling blinding terror in a child. This was my piano
teacher's approach.

Her name was Mrs. Karp. She was in her sixties, possessed of
a martial bearing, spoke in a bark-like German contralto, and
suffered no imperfections in her little charges. My parents were
delighted; they praise her to this day. In my view, if that woman
could not beat the natural love of music out of a child, I don't
know who could. In the five years I spent under her tutelage,
my most cherished hope was that someday it would all end. No
means was out of bounds if I could bring the desired result. I
feigned illness, threw tantrums, demeaned myself by begging,

and when it became clear that my parents were immovable, I sought spiritual intervention in the vain hope that a loving creator could provide a painful and permanent solution to my woes. Some children are still taught the power of prayer. Parents might want to warn them of prayer's possible abuses.

All her students were expected to practice every day for one hour. This seems a reasonable enough thing to ask of a child, except when dinner is withheld pending practice's completion. At least all four of us children suffered identically, so there was a touch of foxhole camaraderie in our mutual misery.

At ten I learned that we were going to move to a new city in the fall. I had spent my entire childhood in Pittsburgh, had freely walked my neighborhood streets, knew its museums and parks, and loved my first friends. This sad news set me into a melancholic mood. For reasons I don't ever expect to understand, I decided I would give my piano teacher the gift of learning to play Mozart's Sonata in C major proficiently. I worked at it daily, playing and replaying phrases until they were melodic and meaningful to me. For the first time in my study, I tried to bring feeling to the passages, soft and sad here, strident and insistent there. My playing was transformed.

My teacher was more surprised than anyone. She entered me into the rounds of the local student piano competition, and I exited the winner. So I was advanced to the regional competition in Cincinnati. It was going to be an important affair. There was a celebrity guest judge, Mr. André Watts, who somehow found room in his heart and his demanding concert schedule to listen to a few kids play what they could. I happened to see him on one of those daytime talk shows a few weeks before the competition started. He wore a dark turtleneck and light blazer like an early seventies movie star. He was young, handsome, and theatrical; his playing was explosive. Mr. Watts was one of the few people alive who could convince a kid my age that playing classical piano was cool.

My parents must have been pleased because it was decided that my mother and I would take an airplane trip to see where the musical chips would fall. Cincinnati must be all of forty-five minutes by plane from Pittsburgh. With checking in and taxi-ing, it's probably a faster drive. I surmise that this was a special treat, just for me, my first airplane ride!

It was special indeed. Unless you've seen it, you couldn't know that clouds are piled like mountains, stretch out like fields, and separate to reveal vast acreages of land below, land that looks nothing like it does when you're standing on it. I recommend air travel to everyone. It makes the world fresh, no matter how advanced your years.

I only remember two moments from the competition. The first was sitting backstage in a line with eight other children my age. A distractingly pretty girl sat next to me in the most fash-ionable dress I'd ever seen. I was transfixed. She turned to me, drawing me in with her gaze. Her lips parted. "Is that a clip-on tie?" she asked softly. She was a fiend.

The second moment came as my mother and I walked the path leading from the auditorium. My second-place statuette hung loosely at my side. My backstage acquaintance had claimed the winner's crown. I wasn't upset; up until that morning, no one would have guessed that there was a virtuosic bone in my body. The door from which we left swung open as we walked away, and one of the judges came loping toward us. "Mrs. Elli-son, there has been a change. Mr. Watts insists that your son and the young lady share the first-place award."

My mother and I went back inside. My trophy was exchanged

for something marginally grander. And since that day, that girl and I have been bound, she, triumphant for an hour, and I, the last-minute equal of the devil herself.

Even now, when I sit at a piano, I must overcome a queasy feeling of dread and hunger before I can enjoy playing.

Practice had changed for me. Until then, as is the case for most kids, it had been foisted on me, unasked for and unwelcome. A parade of pursuits followed: sewing, knitting, embroidery, ceramics, bicycle repair, banjo, guitar, drawing, wilderness adventure, sculpture, songwriting, until I was finally launched into the world and had to pay my own way. Carpentry eventually became the first avocation that I thought I could turn into a career. I studied and practiced it with zeal. I found everything about it interesting. Geometry, engineering, craftsmanship, tools, methodologies—they all held secrets I was eager to uncover. My practice brought with it a sense of accomplishment. Every day, I could see the progress I had made.

I was working at my trade but paying little attention to myself. If I slept in and arrived late, it bothered me little. If a job ran too long, that was my boss's concern. It took a few years for me to consider applying the rules of practice to my character. The going has been slow.

Three years into my carpentry career, I was a budding craftsman, full of myself in my quickness to learn new things, and overimaginative in my assessment of my abilities. Because I could hang doors, put up moldings, and generally deposit things in their correct locations, I thought of myself as skilled. It's amusing, and a dash painful, to look back at how little I knew, and how much I did stupidly back then. I was interested, useful, willing, and careful at best. But, to my credit, these are the beginnings of competence.

Despite my willingness to work hard, like most twenty-one-year-olds, I had a lot to learn about deadlines. If things got done whenever we got around to them, that was fine with me. I had not yet encountered the idea of doing everything within my power to meet a goal. Sure, I'd displayed some rigor here and there, but nothing sustained, and nothing that someone would mistake for a personal standard.

I worked for an outfit that would do simple renovations around town. My boss had me build things I shouldn't have, messing with wiring, plumbing, tiles, carpentry, whatever the job required. All of it was exciting to me, and most of the results were professional-ish. Our customers didn't complain. My manners were good, and we came pretty cheap.

Early one summer, a friend of my boss's asked if he could borrow me for a while to help him fabricate and install kitchen cabinets in a suburban beach home. The man was up against a hard deadline; he and his family were heading out west in three weeks, and the job needed to be complete before they left. All was agreeable to my boss, so the next Monday, I found myself in a well-appointed, old-fashioned woodworking shop on the eighth floor of a building in Manhattan's fashion district. The machines were large and worn; many of them I had never encountered before, but I liked being near them, hearing them run, feeling their heavy, even vibrations through the palms of my hands.

My temporary boss was a rangy, intense man, funny at times, but with a distracted, dislocated edge. He would tell me what was to be done, show me the basics of the technique needed to do it, and disappear for the rest of the day. In two weeks' time, what remained of the cabinetmaking was complete. I hadn't ruined too much, maybe two sheets of Formica, but I fretted over it painfully, and I lied about what happened, amplifying my agitation. The truth would have been a better choice; a little disapproval is less disturbing than the terror of discovery that lying grows in the gut.

We wrapped and packed the cabinets, loaded them into a rental truck, and headed for Long Beach, Long Island. Temporary Boss drove the rental; I followed in Regular Boss's Civic, which he had loaned me mostly because he hated finding parking for it near his Upper West Side apartment.

The jobsite was a nicely kept bungalow beach house, built close to others of the same vintage, all varying delightfully in style and color. The neighborhood had an Old New York feel about it, built by people whose kids used terms like "moxie" and "cut a rug." I would have liked to explore it. I never got the chance. The installation started that day, and for seven days, it never stopped. Temporary Boss hadn't said a thing about the work schedule, only that we needed to be done in one week. I'd never installed a kitchen before, so I had no way of estimating what the job would take.

It was a large eat-in kitchen, taking up most of the first floor. There were upper cabinets, lower cabinets, a pantry, a curved seating table, banquettes, counters, appliances, and plumbing fixtures to put in. The design was early eighties Formica sleek. Everything aligned with everything so that each cut, fit, and scribe had to be right on the money. I was entirely out of my depth, but Temporary Boss seemed to have a handle on things. He continued to teach me the technique of the day and set me loose to do my work unharassed. Only now, he didn't leave. No one left the house for the rest of the week, except to get food from the corner deli. The first five days we worked for eighteen hours and slept in packing blankets on the floor. The sixth day we worked twenty hours, catching catnaps every eight hours, and breaking only for meals and coffee. The last day, we didn't sleep at all. We clocked 134 hours that week. With time and a half for overtime, I was paid for 181; there aren't that many hours in a week.

The funny thing about the week was that nothing about it really bothered me. I became fast friends with Temporary Boss. He taught me enough new skills that I could almost have installed a kitchen myself by the end of it. We slept on hardwood, ate bad deli food, drank soda, coffee, and seltzer, and became disorientingly tired, but there was camaraderie and a clear goal in the work that made it worthwhile. Finishing on time felt heroic.

Still, at the end of the last day, I was a mess. I was wearing the same Sears coverall I had arrived in, except now the arms and legs had been hacked short, and it had only been laundered once

or twice in the sink, then worn until dry. I hadn't washed much or shaved at all. I'd lost about ten pounds. There was thirty cents in my pocket; all my cash had gone to the deli and all I had from Temporary Boss was a personal check, the largest I'd ever seen. It was time to drive back to the city, and I was dizzy with exhaustion.

Temporary Boss climbed behind the wheel of the rental truck; I fell in behind him in Regular Boss's Civic. We drove as far as the nearest gas station and pulled in to fill up. He paid for both our tanks and we set off for the city and sleep. A few miles down the road we came to the bridge that connects Long Beach Island to Queens. Temporary Boss stopped at the toll booth, paid, and went through. I pulled up and looked hopelessly at the sign: $1.25. I had thirty cents. The man at the toll booth scowled at me, made me pull over, checked my license, took my address, gave me a short lecture, and sent me on my way. Temporary Boss was long gone.

I didn't know my way back to the city, so I used the sun to guess the direction and kept taking turns to correct my course. I went from road to avenue to on-ramp and finally found myself on an unfamiliar highway that looked promising. A few miles farther along two signs loomed overhead, one reading NEW YORK CITY, the other NEW ENGLAND. Some unpredicted synapse fired in my brain: I repeated the sign's text out loud in a singsong voice, "New England," and steered the Civic that way.

I drove in a sort of dream for an hour or so, not thinking or planning or worrying where I was going. I crossed into Connecticut with plenty of gas and nary a care. A few exits later, another sign appeared, KENT, EXIT ½ MILE. "Judy lives in Kent," I intoned aloud. Seventeen-year-old Judy reconstituted herself from my mental vapors, right there in the Civic. Some apparitions chill a man to the marrow; this one brought a warm, sweet glow. She was the one bright spot from my last attempt at schooling a couple years earlier. We had separated in the way classmates usually do, but we had never properly broken up. Surely, she would be delighted to see me again.

Another few turns and I was at her door. I rang, and knocked, and rang again. The door cracked open a smidge, and then swung wide. "Well, don't you look like something," she proclaimed with a smirk. She fed me soup and sent me first to the shower, then straight to bed. I slept for twenty hours.

It was mid-morning when I woke up. Judy made eggs. We caught up on the events of the years since I'd last seen her. She had an easy way about her; she could talk with anyone about anything and things with her were always fine. She never even asked what I was doing there. "What would you like to do?" That was her kind of question. "Let's drive around," I offered. So we hopped in the car and took a tour of the town. They had a branch of my bank, so I deposited Temporary Boss's check. Some of it cleared right away; I withdrew the cash, and we ran around together for the day seeing how we could spend it.

The next morning, we woke up and Judy announced that she wanted to drive to a local waterfall. It was a pretty place. A steel bridge crossed the river just downstream of the falls; the water rushed under it making a singing sound against the steel. We parked at the end of the bridge and walked to the middle for the best view. Judy and I chatted and admired the falls; I've always been mesmerized by running water. I twirled the key chain absentmindedly around my finger. Without registering what was happening, I watched the key chain slip from my hand, fall through the grating of the bridge, and make a tiny splash as it fell into the rapids below. "Fuck," I said. We were stranded.

Judy was pretty, so we didn't have much trouble flagging a ride back to her place, but the damned car was at the bridge with no way to drive it. Regular Boss had expected it back two days ago, so about now it was probably officially stolen.

I was already in Dutch, so I hatched a plan that made sense to me at that age. I knew where Regular Boss hid the keys to his apartment door, and I knew there was another key to the car in the drawer of his desk. I ordered a taxi ride to the train station and bought a ticket for the city. I loitered my way to the entrance of my boss's building, found his keys in the crook of the pediment where they always were, and went upstairs to get the car key. It, too, was right where it should be. I glowed with my success. The day had the feeling of a secret mission: I had found my way in, gotten the goods, and left without a trace. Stopping at a nearby bank, I took out all the cleared cash and hustled onto the train back to Judy. Another cab took us from her house to where the Civic was still parked. Now we had money and wheels, all two silly kids ever needed.

Five hours later, we were three hundred miles away. For the next ten days, we swam and cooked and canoed at a lakeside cabin in the Adirondacks. No phone, no electricity, no shower, just fun, food, and care-free frolicking. But we couldn't stay forever. Judy had to work, and the money was running out. So we hopped in the now certainly stolen car and drove back to her place in Connecticut. A pang or two of guilt had arisen in my midsection. I still wasn't sharp about these things, but it seemed like I should drive back to New York City, my road not taken two weeks before. I kissed Judy goodbye, thanked her for a wonderful time, and headed to the Upper West Side. A precious wink was the last I ever saw of her.

Regular Boss was home; he was loud and full of questions when he heard my voice through the intercom. He buzzed me in. I climbed the four flights of stairs to his place, not knowing what I was going to tell him. "What the fuck, man?" I still couldn't think of anything. "I called my friend. He thought you might have fallen asleep and crashed somewhere." "No, I went to the Adirondacks." "For two weeks?" "Yeah." "You stole my car." "Yeah." "What the fuck? Are you okay?" "Yeah." "We thought you were dead." "No, just tired." "You stole my fucking car." "Yeah, I'm sorry." "Don't steal my car again." "I won't." "You sure you're okay?" "Yeah, I'm fine."

"You're a dick. Go home. Meet me here Monday; we have a new job."

THE PRACTICE OF PERSUASION

Other people's motivations are invisible. Some can be inferred. Parents, if they are decent, are concerned with their children's safety, well-being, and success. Bosses are generally in the business of making profits and controlling costs. Lovers are often shamed into hiding what they want for fear of rejection and ridicule. Likewise, my motivations are invisible to others. It's unreasonable to expect anyone to guess them, and inference is a notoriously inaccurate tool. There's something I've tried to teach my three sons that now seems so obvious and plain that it was incomprehensible to me for a long time.

Ask for what you want. No one is likely to offer it.

It's the simplest idea in the world, but if you apply it in a few common settings, the trepidation of doing so becomes apparent:

1. "I want to drop out of high school."
2. "I love the smell of your hands; may I taste them?"
3. "My Christmas bonus was half what I want; please double it."

These are all examples from my own life. To my great surprise, all three were met with positive responses. I have learned that it might just be the asking that takes more practice than anything else.

I had been a conventional carpenter for a dozen years before my first real technical challenge presented itself. It was a round entry hall, in my first billionaire's penthouse, which was to get an elliptical domed ceiling in traditional three-coat plaster. It took a day or two of pondering before I hit on a process that I thought would work.

I decided to fashion a radial array of quarter-ellipse ribs for the ceiling's structure; then I milled custom angled borders from PVC tubing for the constellation of recessed lights that ran-

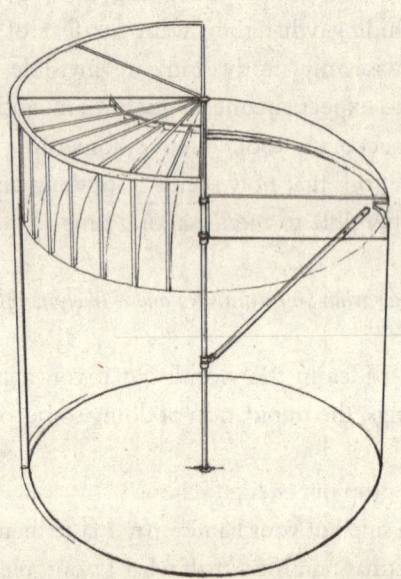

domly dotted the dome. Finally, I set up a central stanchion made of steel pipe to spin plastering knives around for the arc of the ceiling and for a crown molding where it met the wall, like an inside-out potter's wheel. I let the plasterers do the rest. I enjoyed the work and learned a good deal. Only later did I find out that I had done it in half the time that the contractor had budgeted, and without the helper he had supposed I would need. Of course, he wasn't the one who told me; contractors need to make profits where they can. His estimator, who always said too much and had no share in profits, let it slip.

After building a few more curvy things, I was assigned a stack of doors to install. They were eight-foot doors, to be hung on offset pivot hinges, in flush jambs that were stacked, unassembled, next to them on the floor. It was the first time I had ever been given more than two doors at a time to install, so I set about applying the advancements of Mr. Ford, and turned myself into an assembly line. The first day I carefully surveyed each location and made a list of measurements reflecting the conditions I found. The second day, I cut each jamb leg to size at an angle matching the floor and began routing all the hinge mortises. On the morning of the third day, my foreman, Tony, came by and asked what the trouble was; I hadn't hung a single door. "I'm mass-producing them," I said. He was used to my impertinence and trusted me enough by now. "Okay, have at it!" He half saluted and went on his way.

That day I routed all the mortises in the doors and screwed the hinges into place. On day four, I set all the jambs and, by its end, got my first door up. The fifth day I hung the balance of the doors by lunch and spent the afternoon planing and fine-tuning little inconsistencies in each opening.

Monday morning came; one of the general contracting partners, the loose cannon of the two, presented himself where I was working. He was not a talker, more of a blusterer: "I hear you can hang five of these doors in a day? Most of these idiots can only finish one!" I did the math in my head; I had hung twenty doors in a full week. "No, it's four." "What's four?" "It's four doors. I hung twenty doors in five days. I can hang four a day." "Goddamn. That's great!" He spun gracelessly on his heel and exited. A few days later I was assigned to one of his projects.

Christmas arrived on his job. I got a week's pay as a bonus. I had worked for the company for two years without a raise, and I made exactly what every other carpenter in the company made. This annoyed me. As work was ending one evening, I went to Blustering Boss's site office and knocked on the door. "Yeah!" he answered. I walked to his desk. "I've been here for two years. I outwork every carpenter you have. I want a twenty percent raise." His bluster turned to sputter. "We're on a fixed contract here with the owner. I have to bill every carpenter at the same rate. Once this project is done, we can talk about it again." In my judgment, and I'm routinely optimistic, there was at least another year to go on his project. I left without saying much. I still can't believe he didn't go for it, but I hadn't properly prepared my math, and I was not as practiced at audacity as I am now, so his math prevailed.

Blustering Boss's project was filled with elaborate woodwork. I spent the next two months befriending the owner of the company that made the nicest of it and took his job offer as soon as it came. I asked for 40 percent more than I was making and got it.

Twelve years older than me, my new cabinetmaker boss lived in an unfinished house near a rocky northern coastline and had built himself a dojo in his backyard. Given to calling people

"Man," he was a holdout hippie, having grown up in a time when dropping out of school and taking on a craft were de rigueur. He had a motorcycle-riding foreman who was the talent in the operation. That man lived and breathed woodwork; he was fantastic at it. Hippie Boss was content with the role of "spiritual adviser."

The foreman and I formed a fast bond; he liked the way I worked and trusted me completely with the field side of things. He would send shop drawings to me by courier; I would measure the site in question, carefully mark the drawings with my corrections, and truckloads of millwork would show up for me to install without the two of us ever speaking. I never sent back a stick of it.

Hippie Boss was into motivation, so he offered me a sizable bonus if I could finish my first large project on time. Unfortunately, Hippie Boss wasn't much of a businessman. It wasn't long before he started complaining that I made more money than he did. Every few months, he would send his neighbor's nephew or a shopworker's cousin to join me, asking that I "show him the ropes" in the hope that I would train my own cheaper replacement.

I'd befriend the new fellow, who was a long way from home, tell him where the sketchiest "gentlemen's club" in town could be found, and feign dismay the next few mornings when Hippie Boss would call to ask how the new guy was working out. All I could say was that I didn't know; he hadn't shown up yet.

After two years of this, several successful projects, four failed replacement installers, no bonuses, and one tragic motorcycle accident, I'd had enough.

My next job was more of the same. Another bonus was dangled and forgotten. After a year, I got a short telephone call explaining that I was now expected to pay disability insurance in arrears for my time there. These things are hard to countenance with a wife and three kids at home. I quit millwork installation during that call and with a few more calls found a job as a site supervisor. It paid 20 percent more than installing and was refreshing in its way. The broader scope of it appealed to me. Wood is beautiful, but all day, every day, it loses some of its luster.

I worked as a site supervisor for a few years until the economy faltered and the industry tanked. Construction in New York City is the caboose of the stock market's toy train. A recession in the early nineties restricted the profits that could be leeched from the world's enterprises. Those who depend on economic expansion felt the pinch, decided against lifestyle upgrades, and a few years later half my colleagues found themselves with nothing to build.

The company I was working for ran out of work, too, so I went back to installing for a local cabinetmaker who had just landed a large project with one of the city's most prestigious general contractors. A shaky security can be found in working for people who are unaffected by the ups and downs most of us endure. In times of hardship, I retreat to the companies that build for that set. The richest among us are always building something somewhere.

My new cabinetmaker boss sent me to a five-thousand-square-foot apartment on Madison Avenue a few blocks from the Metropolitan Museum of Art. It was being gussied up for the aging heiress to a soup-making fortune. We had been hired to install everything wooden except the floors: doors, jambs, casing, baseboard, crown, paneled sitting room, kitchen, vanities, closet interiors. My helper, Donnie, and I were the only millwork installers on the job, but we were more than up to the task.

The general contractor in charge of the overall project took a keen interest in my work. He had never seen carpentry systematized. After he'd gotten to know me better, he would ask streams of questions about how to bring the approach to his whole organization. Over the years, I had expanded the one-man assembly-line method to most everything I did. It worked so well that I had grown impressed with myself. I fancied that I might be able to show a few crews a thing or two.

But this contractor and I hadn't gotten off on the right foot. The first time I met him, he had received a call from his site superintendent complaining that I was unkind. I suppose I had been. When Donnie and I arrived at the appointed time on our first day there, with a truckload of pre-finished floor-to-ceiling paneling for the sitting room, we found the space filled to within two feet of the walls, and three feet high, with the project's accumulated debris. There was no place to put our delivery. I waited the better part of the day for the super to have it cleared out. When annoyed, especially justifiably, I am unable to mask it. Site supers are theoretically the top dogs on their jobsites. They rarely take kindly to a heated ear chewing.

The general contractor was none too pleased with his super's call. He, in turn, called my cabinetmaker boss and lambasted him regarding my unprofessional manners, only to learn that my boss found my manners both appropriate and amusing. Unsatisfied with this outcome, the general contractor decided to visit the site so that he could lambaste me in person.

During our first encounter, I was ten feet in the air on a ladder nailing the pre-joined inside corner of a lacquered crown molding to the walls. "Is that crown coped or mitered?" he asked. His tone was snarky. I had no idea who he was. He was nattily dressed, with a pricey haircut. He carried a man purse, so I surmised "management." "If you can't tell from there, what difference does it make?" My comment didn't faze him. He walked closer and put his index finger into a hole in my shirt. "You might want to get that fixed."

"What makes you think comments on my clothing are welcome?" I hissed. We were off to a great start.

The contractor dropped the conversation and left me to my work. My contempt slid an inch toward respect.

This sort of project had become my carpentry sweet spot. Donnie and I breezed through the installation without a hitch, and I even repaired my relationship with the super I had offended at the outset. I still have an amusing photograph of Donnie standing in front of the Van Gogh that hung over the sitting room fireplace, a painting worth more than the entire crew's combined lifetime incomes.

On the day we were packing our tools to go, the general contractor and I found ourselves sitting on some church steps nearby on Madison Avenue. It was a breezy summer day. He looked at me crookedly. "How do you figure that much money? That's three times what I pay my other carpenters." "Well, the way I look at it, I do the work of at least three of your guys, but since you also want me to take on some supervision, I'll just ask for the carpentry money and throw in the management for free to make it worth your while."

It pays to do the math before any negotiation; that way the boss knows that when you ask for a raise, you've already earned it. It also pays to throw in a bit of an incentive. And if anyone mentions a bonus, ignore it; if they want to give it to you, they will. I was hired on the spot.

From that day on, I have generally remained satisfied with the money I've earned, but no one has ever, of their own volition, offered me a raise; I've asked for every one of them. I like it best when companies have policies requiring yearly employee review. The silence on the other end of the line is precious when I call the boss and say, "It's my hiring anniversary!"

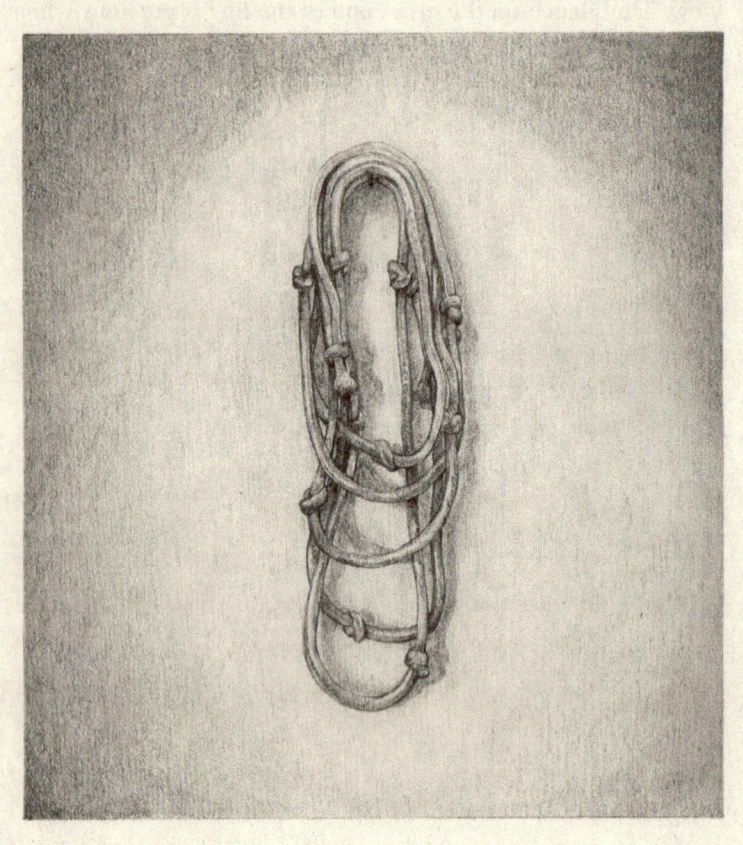

CHAPTER 4

Math and Language

Math is a blade; language is a cudgel.

MATH

The world is utterly abstract. We see it by reflected light and hear it through tiny distorting tunnels. Everything comes to us in bits and scraps, always somewhat discolored by our very means of perceiving it. Hoot owls and hyenas live in far richer sensory worlds than ours. Yet, somehow, looking out through a few smudged windows at the arrayed universe, humans have deduced worlds within worlds of particles, energies, forces, motions, and processes. If science is to be believed, an idea that at least deserves entertaining, each of us is a walking galaxy of madly swirling bits of energetic matter, spinning so fast that when I reach for my coffee, my hand doesn't pass through the mug but grasps it, even though both are made mostly of nothing. How is anyone supposed to think about these things, let alone transmit them to the unbelieving? For this purpose, we have been given Math.

Math is a tool well suited to every manner of ends. When I encounter someone who doesn't know its uses, even at the relatively simple level our work requires, my heart sinks. Everyone's math should be at least good enough that they can take measure of the world they inhabit.

Math's beginnings were doubtlessly humble. When I was schooled, we were told stories of nomadic shepherds who kept pebbles in hide purses they hung from their belts. If each pebble could be matched to a sheep at the end of the day, all was well. Therein lies the beauty of Math: It works. Geometers calculated the curvature of Earth; Galileo reportedly dropped a marble and a cannonball from Pisa's tower, pleased with himself when they landed together; Newton boiled it all down to a few simple laws; and an optimistic intellectual elite was seduced into the sunny belief that human beings can figure out anything. They weren't entirely deluded. They looked out, they looked in, they measured, they calculated, and much of the world was made comprehensible.

Then, in 1931, following the impulse of so many of his contemporaries, Gödel blew the whole thing to pieces. I can only imagine the mathematicians of his time working their way through the proofs of his incompleteness theorems and muttering, "Shit, No, Shit, Aaargh, Fuck, Fuck, Fuck!" as their whole world dissolved before their eyes. He had proven that nothing is provable. Math had turned on itself and slit its own throat.

Decades of sulking ensued. Depressing new philosophies were invented.

People love to overdramatize.

There is a subset of woodworkers that is obsessed with sharpening. I've witnessed heated arguments about the relative merits of Japanese water stones versus diamond hones, buffing versus stropping, on and on they prattle. I can't bring myself to care. If my tool's edge can find its way cleanly through the wood I'm working, and without too much effort, I'm pleased. To me, the point is to make things, not to spend hours perfecting the tool that helps me do that. No matter how sharp an edge is honed, under a microscope it appears gullied and craggy. All methods are imperfect. Nothing is ever as sharp as we might wish it would be. I am happy to take a good enough chisel, sharpen it in a reliable, repeatable way, and get on with my work. Someone out there is waiting for their staircase. It won't be a perfect staircase. It will be riddled with tiny errors, unnoticeable cheats, subtle compromises, no matter how far I go dividing inches into fractions while I make it. But made well enough, it will serve its purpose for years and someone may even grow to cherish it.

People think I'm a genius because I remember my high school math. Algebra, binomials, geometry, trig—these are the everyday language of building. Most people think SOHCAHTOA is a secret code, like the mysteriously carved CROATOAN from that Roanoke tree. They inhabit the same mental space, filed under "High School/FORGET." But used well, that enigmatic mnemonic acronym can be a marvelous key that gets us out of all manner of pickles.

Like all languages, math can be used in as many ways as there are people who understand it. Math speaks of location and relationship as musical notation speaks of pitch and harmony. The world has had Bachs and Beethovens who used their language with the same ease and beauty with which Newton used his. Before them, I gape, I marvel, and I misunderstand. My math is like country music, three planes, and the truth. It gets me through the day, and the occasional long, lonely night. Gödel may have blasted the universe to bits, but my world carries on just fine with Euclid's pretty proofs.

Builders are a resourceful lot. Even without formal schooling, they have come up with countless ways to assemble things accurately, sometimes with homespun tricks that might impress a seasoned geometer. It also helps that, as long as we're building within earshot, Euclid has things nicely worked out. Quantum mechanics might keep me up of a night, but it never makes me second-guess where to place a window.

It's said that the builders who laid out the Gothic cathedrals did much of their work with a knotted chord divided into twelve equal segments. It's likely that journeyman masons and framers knew how to use this simple measuring tool in an endless variety of ways: laying out right angles and arches, dividing spaces into two, three, four, or six equal parts. A pair of clever twelve-year-olds could find fifty uses for it in a day's play. It's just as unlikely that those workmen had any more idea of the mathematics behind what they were doing than those twelve-year-olds would. Perhaps a scholarly brotherhood or secret guild was responsible for guarding that knowledge. Walking around a Gothic cathedral, or taking measured strides of its floor, you can begin to see the twelve-sectioned, thirteen-knotted rope everywhere, arcing through the vaults, triangulating the columns, dividing the tracery at the windows.

A BEAUTIFUL MIND (FROM CERTAIN ANGLES)

Every so often an architect comes along with a bona fide vision. Most are content to study a smattering of architectural history, read a manifesto or two, cower through a series of charrettes run by over-lauded professors with envy-forged axes to grind, then promptly attach themselves to whatever current design trend brings them customers with the fattest purses. The few who survive the wringers of credential getting and client landing still clinging to their aspirations come as a warm breeze in winter.

Leaf through a few contemporary design periodicals some evening and you will see that there is an overwhelming preponderance these days of neatly arranged rectangles in five shades: black, white, silvery metal, wood/beige, and gray. Here and there some subversive upstart will swirl a few curves through a room, or paint a study in jewel tones, but we are noting a trend at the moment, not highlighting its detractors. It's as if some focus group of potential interior design customers was assembled and asked to list their preferences. Since those five shades and rectangles appeared on everyone's list, architecture's governing body, with a grand name like "The Prefecture," declared that henceforward all rooms, everywhere, were to contain these six fundamental elements.

What horn-rimmed revolutionary wouldn't want to smash the whole infernal order to bits? Now and then, one will try, but like many an escaped jailbird, most find out quickly that it helps to have a well-thought-out plan, especially now that the authorities are after you.

Beginning twenty years ago, I worked with one architect through three successive projects. He squirmed and struggled as he was dragged hither and thither by the demands of a triumvirate of clients who cared little for his theories. He was thwarted at many a turn. None of the jobs panned out as he had envisioned them. After five years of these indignities, we went six years without crossing paths.

By pure happenstance, those six years later, I found myself

sitting in a contractor's conference room, negotiating a position I didn't want in a company I didn't want to join, when our talks were interrupted by a young manager who barged in and asked, "What do you want to do about the Hotson staircase?" My ears pricked up. "David Hotson?"

"Yeah, you know him?"

"Sure, we did a bunch of jobs together. Do you have drawings?"

Indeed, they did. They rolled them out in front of me on the conference table. "Look at that!" I exclaimed. I was so proud of him. There before me lay the most cockamamie tangle of architectural explication I had ever seen. Lines went every which way, connecting at apexes scattered willy-nilly around the page. A vaguely helical worm ate its way through the whole structure, as if it were the Serpent's own apple. "Let me take these home. I'll send you a price in a week."

One thing followed another, and in a half year's time, I found myself in a four-story penthouse atop a lovely old publishing building. From the catwalk encircling its roof, half of Manhattan's legendary structures could be seen close at hand: the Brooklyn Bridge, the Woolworth Building, the empty footprints of the World Trade Center. We were close enough that we could make out the details of the gold-covered statue atop the Manhattan Municipal Building, improbably named *Civic Fame,* as though it were a classical Virtue.

I busied myself locating apexes with my plumb bob, laser, and lengths of strong fine string called Jetline. I had already built a one-eighth scale model of the complex central stairwell, and a full-size mock-up of the most demanding intersection within it. I could find no existing combination of materials and methods made for such geometric eccentricity, so one invented itself in my head as I strolled through the composite decking section of a Home Depot one afternoon. I was as surprised as anyone by how well it worked. Cheap fake-grained vinyl deck boards served extraordinarily well as the custom-cut edgings that formed the folded corners that ran throughout the structure. Walls met at the most unlikely angles, refracting the surrounding space into a collection of vistas, revealing the building's riveted steel skeleton and incomparable city views beyond.

My haphazardly assembled crew consisted of Beth, a former assistant project manager who was game to tackle the messier world of the site; my oldest son, Matt; and Darren, Fynn, and Michael, three freelance carpenters who had made the rounds of upper-echelon New York building. They showed uncommon patience and devotion to the job, between them setting more than two linear miles of vinyl edges along carefully laid Jetlines.

The penthouse's owners were as kind and involved a pair as I have ever found in this business. Both inhabited the highest reaches of the tech world. Who knows the schooling, years of work, and unrestrained effort they must have put forth to reach such levels in those companies? The husband found my crew amusing. He would come around asking how we did what we were doing and expressed genuine admiration at the results. Even when I tore off on unsolicited tangents, about how the metric system makes children into humdrum finger-counting simpletons, or how advertising robs us of knowing what it is we might individually want in life, he would listen with apparent interest, even asking a probing question or two to further the discussion.

One morning, he found me halfway up a flight of steel stairs that we'd installed to support the finished stairway of white opaque glass that was to come. I had written in thick black marker a series of equations calculating the exact height of each glass tread. Years earlier, a cabinet shop foreman had shown me a simple trick for keeping track of increments smaller than the $\frac{1}{16}$" marking where most tape measures give up trying for accuracy. He would draw a little arrow up or down after the readable measurement and then put the number 32 in a circle, indicating, for instance, the more exact measurement was $7\frac{11}{16}$ ↑ 32.

Now you could call this $7\frac{23}{32}$", though few carpenters could readily find it on a tape measure, but anyone with a little experience would understand that this was midway between the $7\frac{11}{16}$" marking and the $7\frac{3}{4}$" marking. I found the system clever and later expanded it to include little circled 64s and 128s. I would tag these on to a measurement that fell between the tape measure's markings with an arrow up or down, and now I could measure to a much greater degree of accuracy and still easily find that measurement on my tape. Over the years I learned how to arrange these simple hieroglyphs in long columns with the arrows and circles to the right. I could add, subtract, multiply, and divide with all the accuracy I ever needed for building. My client looked quizzically at my columns, arrows, circles, and calculations, which now covered an entire wall. This was a mathematics he had never encountered! He ran to grab his camera. He snapped a few pictures of me and my "formulas" and has prized them since, calling them my *Beautiful Mind* moment.

This is flattering, but it is also fatuity. The work of real mathematicians like John Nash is so far beyond my understanding that I don't venture into their waters for certainty of drowning. Like Newton seeing gravity in his apple, Nash laid the foundations for game theory and differential geometry by closely watching the seemingly random movements of groups of pigeons. This job's client understands the intricacies of things like *machine learning* and *computational biology*. He can follow Nash's

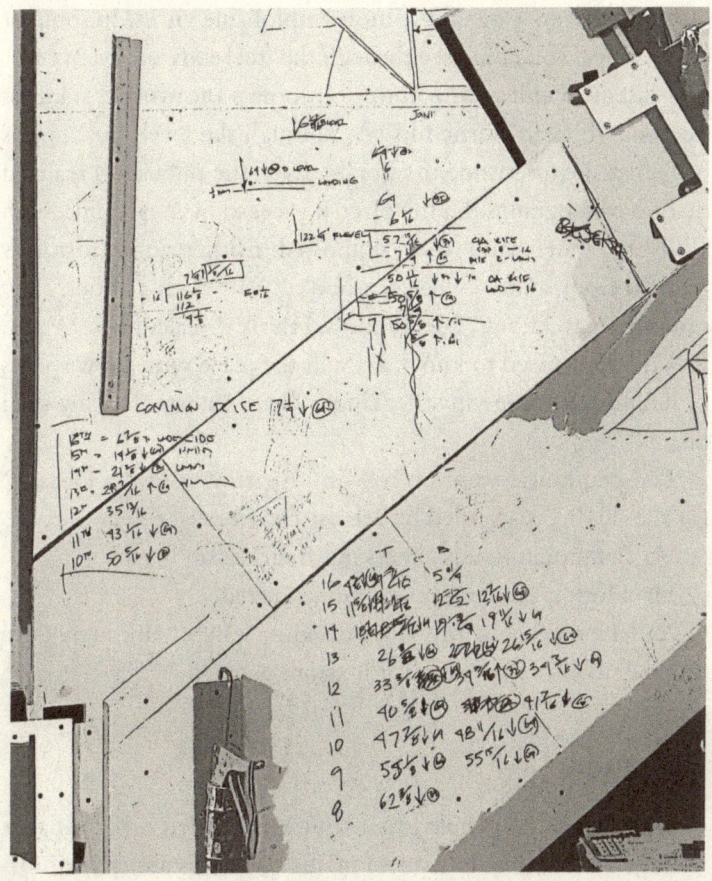

formulas and understand their logic. I can't. All I was doing was simple arithmetic, a slightly hot-rodded version of what most people can do by fifth grade. Still, we built a heck of an impressive stairwell with it, something neither he nor Nash could do.

A few months after we completed the place, my son and I came across an article marveling at the architectural achievement presented therein. The pages were decorated with splashy photos showing some of the more improbable views. In some of the pictures, you could see through the steel sixty or seventy feet to a distant window or painting. In others the wormy stainless steel slide French-horned its way through the levels, making its bell-shaped exit outside the library. Even the publishers realized that no photographs did the place justice, so over each image, in a silvery watermark, they embossed mathematical formulas cribbed from some college textbook. "What's that about?" my son wanted to know. "I don't know. I think it's calculus." "Damn, they'd be surprised to know we built the place with a few spools of Jetline and some vinyl decking!" There's a boy after my own heart.

The penthouse went on to win "Apartment of the Decade" for the 2010s, as awarded by *Interior Design* magazine.

My computer genius clients got their dream home.

Mr. Hotson saw his grand vision realized.

And I proved out my vinyl decking system, although I am unlikely to ever find a use for it again.

LANGUAGE

One of the things people love about Math is that it is verifiable. Within the confines of a given mathematics, a statement or formula can be tested and proven to be true or false. When Einstein first published the field equations leading to his general theory of relativity, several of his contemporaries did just that. In fact, he had gotten the math wrong in a few instances, but once it was corrected, his outlandish and beautiful theory was proven true in the way Math sees Truth. A hundred years later, experimental physicists are still finding ways in which his predictions are correct, and they are still coming across ways that they are not. It doesn't matter if Einstein sounded crazy, or looked crazy, or forgot to tie his shoes; mathematicians know he accurately theorized the answers to some of physics' most confounding questions. As is the way with Science, a whole new crop of questions, equally confounding, sprouted up in their places. No one in Science or Mathematics is the least bit disturbed that, as a species, we don't know everything. It provides both job security and a sense of delight that there is plenty left to figure out. The particular beauty of mathematics is that it demands proof. The only accurate thing "unbelievers" can say is that they don't understand what Einstein gave us, not that he was wrong.

Despite the efforts of fact-checkers, historians, and forensic psychologists, Language has no such tidy provability. The reason for this is simple: People lie constantly. It's probably humanity's most reliable characteristic. Historians know full well that much of their task is to see through the inaccuracies of every first-person account they discover. How does one decide just how much of some great figure's diary was written to minimize embarrassing personal shortcomings, or to make an encounter with an adversary appear just a tad more heroic?

Thirty years ago, having previously displayed no discernible desire to do so, I took up acting for several years. An acting teacher who had recently moved from Los Angeles to New York wanted to make some simple improvements to his modest apartment, and I was chosen to do the work.

As I often did in those days, I took on the project as a side job. Like so many New Yorkers, I had next to no social life at the time. I had planned to spend my twenties in profligate dating and antisocial drinking, but the AIDS epidemic put a stop to that, upending New York's initial promiscuous promise. Work was a safe and relatively fruitful alternative. I was given free hand in his project to design and build within a few simple parameters. I was happy to have the freedom and treated the job with special care. Near the project's end, I even built a sleigh-style trundle bed just because we thought it would look nice.

The acting teacher and I struck up a friendship, and he decided I would be the ideal guinea pig on whom to test his newly synthesized acting method.

He was correct in his assessment. While most of his students had acted in some capacity since childhood, and had developed fixed habits and worse, I had nothing to work with other than whatever instructions he gave me. I was the perfect tabula rasa on which he could inscribe whatever technique he found fitting. He would send me up to the front of the class, have me perform a theatrical snippet, and then give me some specific instruction, such as "Now treat the coffee cup as though your grandmother gave it to you on her deathbed." Lacking anything else to do, I would turn my attention to the coffee cup in question and invest it with an importance that, while completely independent of the playwright's intentions, would imbue the scene with newfound richness and meaning. It's really quite the trick; meaning was made demonstrably independent of the words that were written to convey it. He had a satchel stuffed with such tricks.

After a few years, he decided it was time I went out on professional auditions. I would wait my turn with the hopefuls and do my scene in the way he had taught. I wound up performing in a few forgettable off-off-Broadway productions, and finally abandoned acting for two completely excusable reasons: I still had no desire to be an actor, and the first of my children was born. One can't easily defend committing long hours of unpaid labor to a hobby one cares little for when being the primary provider is one's obviously more pressing duty.

From these years of study, it became clear to me that language is a malleable thing. The words we say are the shaky screen with which we mask our objectives. As authors of our own scripts, we get to choose the details that make our stories palatable to ourselves and edit out the details that don't show us in the best light. Meaning isn't in the words, but in the way we say them, the little sugars and spices we add internally to make our stories tastier to ourselves and to our audiences. "What a cute puppy" can be a pleasant way of easing the tension that arises in the too-close confines of an elevator, or it can be the opening line of an abduction. The words don't matter. When our objectives stray a little too far from what we want our listener to glean about us, words seamlessly slip into lies. Most of the time, no one notices; even if they did, who would point it out when, a moment before, they had done the same thing?

So what's the purpose of talking? Like most inventions, language exposes our shortcomings and weaknesses. We know we're flawed, so we engage in a constant struggle to appear better than we are.

Few people seem to be above the temptation to inch their way up through the ranks of the pack. I say, "I'm sitting on a chair." My companion might counter that it is really more properly a stool. A furniture maker notes that it's a bentwood Aalto stool. A collector sniffs that it's a replica. An engineer highlights the dynamic way in which tension and compression are accounted for in its design. An architect elucidates that it is not really furniture at all, but an extension of the total architectural environment. And we are no longer talking at all.

Still, there is hope in language. There is meaning, transmission, and mutual understanding to be found in Shakespeare's intimate language of mankind's inner world, and in the furtive grunts of chimpanzees. We get tripped up by our intentions.

It seems silly to point out, but language works best when the speaker has something to say that they would like the listener to comprehend. Better still if they have thought the thing through a bit before talking about it. Best of all if the speaker chooses words and phrases that the intended comprehender might grasp, at least in time. The listener of course bears their own responsibility in this arrangement. Silly, sure, except that this is rarely what is occurring.

At fourteen, I left home for good. Both my parents have strings of letters after their names, signs of accomplishment, but more importantly badges of defiance, like prison tattoos for intellectuals. My mother's people were homesteading ranchers from Montana. Her parents got educated and got out. My father's parents were straight-backed Calvinists from a remote corner of upstate New York who took up indistinct roles solid enough to win them a house, a car, and not much more. Education for both my parents meant freedom. It's nice there's something they could agree on. We children were expected to be smart. When we waivered, my father would buttonhole us, point out the failure, and elicit a promise of better performance in the future. It's not a ritual I look back on fondly. When my father sent me a box of childhood mementos a few years ago, my yellowed report cards wound up in the woodstove.

The future came soon enough. Each of us was packed off at fourteen to attend prep school in the belief that it offered the best education to be found. I used to wonder at the luxury of sending middle-class kids to a place designed as a training ground for the stupendously wealthy. But the school was surprisingly affordable due to an overstuffed endowment provided by generations of alumni bent on guaranteeing admission to even their densest progeny. In those years, it would have taken better than half the cost of tuition just to feed me. My parents likely saw it as a good deal: the best education money can buy, and a lot fewer trips to the A&P.

I doubt they understood the bargain we had all entered. Just as no one tells fathers-to-be about the creepier aspects of childbirth, no one tells parents of prep school students that they are relinquishing all authority, for good. For many of the kids there, this was just more of the same; strings of nannies, housekeepers, and doormen had supplanted their parents long ago. For me it was like stretching a rubber band until it breaks. Each new year, I would enter with the determination to do well; I'd find a place on the first trimester's honor roll, and by year's end, I would be failing. When my paternal examination came each June, I could no longer invent a reasonable explanation.

In truth, the school was just as lost as I was. What other school of a thousand students had a classics building at one end of its campus and an audiovisual production lab at the other? Lawns were dotted with smug signs reading ABJURE THE HYPOTENUSE rather than the pedestrian KEEP OFF THE GRASS, while the wealthiest of its denizens dropped acid on those same diagonals.

Two years of coeducation had broadened the pool of applicants, but little progress had been made in managing or even acknowledging the novel challenges a mixed student body engenders. Every student knew someone who had sought out an abortion or was dealing with the mental health fallout that often accompanies one. A prestigious high school couldn't very well advertise that they had the whole pregnancy thing under control. Harder still would it be to point out that they were making steady headway bringing down their faculty/student molestation stats from year to year. The old authoritarian dynamic had become noticeably awkward and hesitant in the face of modernity. Two hundred years of patrician custom was being undone by the newfound realization that adolescents were actual people, not just lumps of moneyed clay.

A few months before my freshman year, an article was published in *Newsweek* magazine entitled "Why Johnny Can't Write." The educational and professional worlds had worked themselves into a tizzy over recent college graduates' inability to string together cogent sentences. This was apparently a shocking development, never mind that the article was simply an echo of Henry Higgins's lament from sixty years earlier.

The august headmaster of my soon-to-be alma mater was singled out in the article for his efforts at undoing the catastrophe. He had introduced a course named Competence that every student was required to pass, and whose values were extended to all corners of the curriculum. We were expected to demonstrate sound reasoning and organized writing in every class—math quizzes were marked down for spelling errors; science papers were inspected for run-on sentences. I credit the approach with my inability to send a text message to this day without proofreading it first.

The results of Competence were telling. A significant portion of the student body suffered through abortions, addiction, and attempted suicides, but most of us left the place with well-established habits of expository fluency and mathematical grounding. Pretty college entrance essays were submitted by applicants who might have spent the summer in rehab. The school's ways were so different from most that the student body had its own accent notable for a distinctive rounding of Ls that I've never heard anywhere else. Perhaps Mr. Higgins could pluck a modern-day flower girl from the city's streets, teach her this unique proclivity of diction, and pass her off as a duchess.

All of prep school's contradictions are hard to square. I'm

grateful that it taught me, a middle-class boy from Pittsburgh, how to write a cogent paragraph, but I remain disturbed that the reason the school exists is to give another boost to kids who already have more than one leg up on everyone else in the world. Perhaps those contradictions are best expressed by the passing remark of a classmate's father.

After high school ended, my sweetheart and I made a farewell tour of graduation parties starting in Maine and ending in Oyster Bay, Long Island. It was a graduation that didn't include me. I had failed a required American history exam and botched the mandatory term paper. It was too near the end for me to care. I sat through my professor's disquisition, which was designed to hammer home that I had "ruined my life," a criminal thing to tell an eighteen-year-old. I suffered the embarrassment of disclosing my failure to my parents and took to the road in a beat-up Volvo with the person I loved best in this world.

No one celebrates rites of passage like the well-heeled. After several nights of drinking too much and sleeping in the car, we arrived on Long Island's North Shore the worse for wear, but content in our freedom. We were put up in the guesthouse of a Gatsby-level estate, fed, washed, and rested in preparation for a night of celebration the equal of which I haven't seen since. My father's tuxedo was ironed and donned; gowns far flouncier than a gymnasium could bear were tugged into place; an orchestra was assembled; great heaps of fruit, flowers, fishy things, and frippery set the tables to groaning. It was grand and beautiful and decadent. I stumbled through dances I didn't know and held the prettiest girl I had ever seen. This is the life so many people want, and I knew full well it wasn't mine.

The next morning, we straggled awake. Breakfasts were fetched, stomachs relined for the trip home. The entire ordeal, all four years of it, was over.

I had my omelet in the company of the man who owned the place. He was relaxed, genial, and wanted to talk. We took to each other immediately. We spoke for an hour, and he walked me and my girlfriend to the car. Before we left, he stopped me with a last remark: "Do you know, you have the rarest provincial accent in the country?" "Really, what's that?" "Upstate New York, educated." "Hmm, no. I never thought about it." And I hadn't. I got in the car thinking how I liked the man. It took me years to realize that I had been insulted.

I worked on an apartment once whose sole design conceit was that everything in it was the same color. Everything: Walls, ceilings, doors, jambs, millwork, and floors were painted and stained in this color. Hardware and plumbing fixtures were powder coated to match. Concrete counters and plaster walls were custom dyed. Sliding glass doors had film interlayers to tint them just so. We carpenters joked that we should have jumpsuits made in matching fabric so that whenever the designers came around, we could zip them up and stand in the corner; no one would ever know we were there. Anyone imagining a placid blue-gray or a vivid orange at this point is far afield. Wherever we looked, our world was putty, that awful lifeless original Macintosh putty. It hurt our hearts to be there.

One morning, when we all wished we had our jumpsuits, the architect, designer, client, and an eminent colorist gathered near me in the master bedroom with three large swatches of fabric. I was halfway up a ladder, and apparently in no need of camouflage to achieve invisibility. The group held up the various swatches, discussing their merits animatedly. Any of these might wind up as the covering of a custom upholstered headboard. Of their choices, two were near misses, and one was a dead match

for the dreaded putty. They each offered their professional opinions, going on at some length. The entire time my internal
voice screamed, "For God's sake, it's the one in the middle!"

Finally, the colorist, clearly the ranking authority in this discussion, took the middle sample and raised it thoughtfully into
the light reflecting off the wall. "I would prefer that we use this
one so that we don't disturb the scintillation austerity of the
room."

And yet again, language was nailed to the cross of intention.

CHAPTER 5

Absurdity

We all hope that our work serves some greater purpose.
Many never find their apex, but like all happily
recovering addicts, I have my nadir.

Sometimes in our endeavors, the stars so completely misalign
that we look for some meaning in their scattering. There must
be a lesson we are meant to be taught, a worm that needs digging out. Surely by the end of it all, there will be an Oedipus or
a Lear wandering blind and ragged, wailing in rage at the fate he
has brought upon himself, the meaning of it all beckoning just
beyond his grasp. It has been nearly thirty years since this tale
took place, and no salving pearl has formed itself around its jagged seed.

THE SNAIL STORY

I was engaged to be married to a woman who was pregnant.
The timing seemed ripe to find a job with health insurance. A
friend recommended me to a project he was working on that
had been under way for more than two years. They needed car-

penters with a creative bent, so it promised to be a good match. I was hired on recommendation alone, no interview, no phone call, just a message relayed via my friend that I was to show up on a Monday morning at the southwest corner of Park Avenue and Fifty-ninth Street, find the freight entrance, and ask to be taken to the penthouse.

There were signs on my first day that things onsite had taken a peculiar turn. The crew was weary and sarcastic. The first carpenter I encountered was blasting Stockhausen on a portable tape player. I remain unfamiliar with his work, but I would agree with the laborer who asked him to turn it off a few days later that its effect was like someone "vomiting in my ears."

For a job that had been actively running for two years, very little appeared to be done. Every room I could see was still in the rough stages; systems were only partway installed; the whole place felt like it had lost its momentum long ago.

The Stockhausen enthusiast told me I could find the foreman on the next floor up; he would know what to do with me. He waved his hand in the direction of the passenger elevators and gave no further instructions. We carpenters do not take passenger elevators, so I searched around and found a fire stair that led up into a short rectangular passage where a silent carpenter was framing new walls with metal studs. The studs were a few inches in front of concrete walls that formed part of the structural elevator core of the building, so he was merely lining an already defined concrete hallway with a finished surface of drywall and trim, a simple task by any measure. Between the studs, scrawled in blue Sharpie, covering every inch of the concrete walls and ceiling behind them, was a dense, meandering maze of words, beginning with the phrase "The hallway . . . shit, I'm still only in the hallway." It was the opening monologue from *Apocalypse Now*, transposed into Carpenterese. The project's architects had drawn this same passage, seen it built, and deemed it unworthy three times already. This poor carpenter had worked in this cell-sized spot for nine months, building and demolishing the same four walls.

My responsibilities at home dictated that I couldn't afford the luxury of trepidation, so I asked where they wanted me and what I could do to help. I was given a room and a task. The foreman shadowed me for a few days to make sure I was up to the work, and by the end of the week, I was a permanent member of the crew. I found my place quickly, happily volunteering to build some of the more ambitious elements of the house. Ambitious it was. Every room was designed by a different artist. It sounds dreamy, but in practice it's akin to asking a clutter of feral cats to guard your sofa.

Most people never really notice that rooms don't bang up against one another. Countless schemes have been devised to wrangle these in-between elements we call "transitions." The importance of transitions cannot be overstated. Like lawyers, preachers, and air traffic controllers, architects can rightly be judged by their ability to manage them. This job had no such guiding light. It was a choir of coloratura sopranos, each singing their most beloved aria at the top of their lungs. Nothing went with anything else: A cherry marquetry sitting room would abut a library with sculptured plaster donuts on the walls, and in between, a half-marquetry/half-donut archway would separate them. Doors wouldn't close; finishes crashed in jarring, unattractive ways; fights ensued.

Besuited project managers cast about for a solution to the chaos. Here is what they hit upon: The blessings of five design professionals would henceforth be required in order to build anything. The artist from room A would confer with the artist from room B in the architect's and decorator's presence. When all came to agreement about a particular transition, they would pass it along to the overarching "architect of record" (a different personage from the plain old "architect"), who would draft the solution and distribute it for all parties' signatures. Then, once said drawing was signed and countersigned, it was to be released to the contractor and us carpenters for execution. Unsurprisingly, the entire job ground to a halt.

Our foreman's girlfriend had recently landed a role on a popular soap opera, so he brought in a portable TV set. We placed it on a worktable in the living room and gathered as a group, roundly cheering her every entrance. We then watched on into the afternoons to gauge the quality of her performance against the work of more seasoned actors on other daytime dramas. The general contractor, our employer, hovered happily by, sending monthly bills for our participation in the "Time and Materials" agreement his contract stipulated. Several weeks of languorous contentment followed, until the day when the owner decided things had gotten out of hand; it was time to clean house.

The owner ran a company with his name in the logo and understood better than most that one person has to be in charge if anything is to get done. He brought in an architectural firm of international renown and gave them the authority to dictate whatever details they saw fit. This stylish, colorless team bustled through the site almost daily, waving drafting scales like bayonets, assuring one and all that things were about to change.

And so things did. The new architectural team was headed by a man whose name was also in the company logo. He deputized each member of his staff, assigned them an area of the apartment, and quickly put the artists into the more fitting role of advisers, reserving anything resembling authority for himself. The change was palpable. The TV was taken away. We all got back to work as the year drew to a close.

On the upper floor of the penthouse, through a wall of glass doors, there was a spacious outdoor deck where nothing had been done in the now nearly three years that the project had been under way. One young architect, possessed of an ambitious energy that separated her from the pack, decided that this would be her bailiwick. It wasn't long before detailed drawings were made, prices were gathered, and components began to arrive.

Here was the plan:

Viewed from the sky, this deck was a rectangle roughly twenty-five feet by thirty feet, with a railing along three of its sides; its western edge was defined by the exterior wall of the penthouse. It was on the thirtieth floor, had a commanding view, and was bathed in sunlight from dawn until mid-afternoon. An irregular grid of stainless steel channels two inches wide and four inches deep was to be arranged over the deck in a sort of tartan pattern. Between the channels, we were to lay thick black slates in semi-regular rectangles, which would be separated by

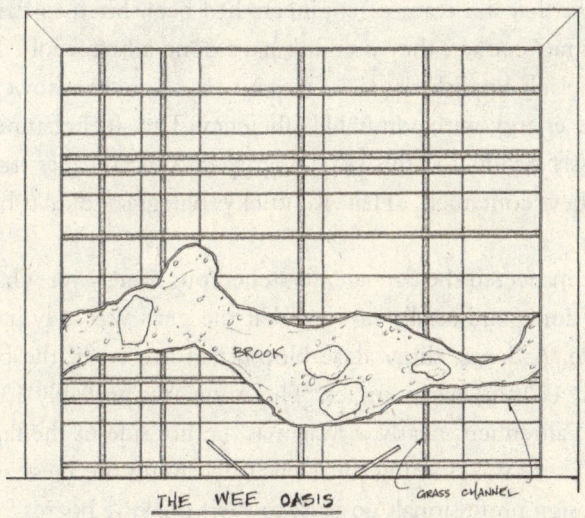

THE WEE OASIS   GRASS CHANNEL

the thinner crisscrossing steel channels, which in turn were to be filled with stripes of soil and planted with grass. Running willy-nilly across the tartan, there was to be a gently flowing, irregular brook, complete with stepping-stones. Presumably, a Highlander/financier would have felt completely at peace.

Spring was upon us, and the work went quickly. In four months' time, the grid was leveled in place, the slates were artfully arranged, Colorado River rocks lined the bottom of the brook, sod was cut into strips and laid into the channels, the pump was started, and the wee oasis was complete.

July in the city can be punishing. Those with means run for the Hamptons. Those without sweat against one another in the subway, stripped to ineffectual minimums, and pray hopelessly that this will be a mild one. No prayers were answered that year; we sweltered.

The morning after the sod was placed, we swung open the doors to the deck to let in any wisp of a breeze. Last night's rich green grass had frizzled to a dust-bowl brown. Thinking back, I warrant that if a team of engineers had been hired to design a solar grass cooker, they couldn't have done a better job. Those thick black English slates, so elegant in a London shower, absorbed energy with admirable efficiency. The steel channels effortlessly conducted this same energy into the meager strips of soil they contained. That Kentucky bluegrass didn't have a prayer.

We inspected the damage, scratched our heads over what was to be done, and finally noticed that the grass was only half the trouble. A dense, slimy algae bloom had overtaken the brook, heavily coating every surface. The water was probably 130 degrees Fahrenheit, nearly as warm as the hot side of the taps inside. There was nothing for it but to clean up the mess and let the design professionals go back to their drawing boards.

On this site, cleaning the upper level was the responsibility of one man, Kevin. He jealously protected his duties, even chiding carpenters for discarding their own clutter, muttering that they were stealing his work. This particular morning, that stance was turned against him.

Kevin was wiry and lined, with a great white beard. He looked like Santa, if Santa had spent three years in a South Polenese POW camp. No one knew much about him. He didn't converse so much as erupt. Holstered at his side was a six-inch hunting knife. He wore heavy steel-toed boots, tied a red bandana around his brow, and whispered about helicopters when he thought he was alone. Through his cussedness, he had earned the assignment of decontaminating the brook. All day he was out there, wielding a plastic scrub brush, pulling the small brown stones from the brook bottom, submerging them in detergent and tossing them into the "clean" bin. By evening, he was done. In someone so generally bothered, it's hard to tell if such a task is irksome. The same pinched scowl we always saw seemed little changed.

Another week went by. A fellow arrived from a swimming pool and outdoor pond outfit. He administered some tests, made some calculations, and announced that our problems could all be solved with proper pH balance and chlorination. The next day he brought a lunchbox-sized device that he installed near the circulation pump, dissolved several scoops of ice-blue powder into the freshly added water, and left until the following morning, when through further testing he determined that his work was complete. Kevin was enlisted to spread his stones in as natural-looking a manner as he could muster, and order was restored to our little biome.

No one was exactly surprised the next day when we opened the doors to the deck and found that the algaeic slime was just as thick and menacing as it had been the last time around. Even Kevin's face was marked more by resignation than ire. He slunk off to find his brush, bucket, and detergent and gamely accepted the inevitable.

Experts were brought in from all manner of disciplines: engineers, botanists, ecologists. A plywood conference table was arranged just inside the glass doors. Refreshments were set out. The entire design and construction management teams were assembled to hear the opinions of the knowledgeable. The slanting sun lent the affair a gravitas it didn't merit. After a few hours, a decision was reached.

Two weeks later, on a Friday morning, a van arrived with two or three unusual crates. They were whisked up the elevator and opened before a small circle of the curious. Customs stickers were affixed to their outsides, and within, sloshing about in clear water, were several hundred smooth brown snails. It was announced that these little gastropods were a special breed. They required almost no care and had a single dietary need: algae. Arroyos in Mexico had remained unclogged for millennia thanks to their work.

The brook was refilled, the Colorado River rocks were decorously strewn about, and the snails were spread throughout, eager to get to work.

The weekend passed. The blistering weather held. We all returned on Monday morning to push our boulders a little farther down the path. We were finishing our coffee and milling about in the morning way when someone opened the doors to the deck. I have come across deer carcasses rotting in the woods and swum among thousands of expired alewives as they washed up on Lake Huron's shores. They were as nothing compared to the fetid air that filled that apartment. Hardened homicide detectives would have blanched at the smell. It was a horrid, life-altering stench. There was no escaping it; it bored into our skins and closed our throats.

The site was evacuated.

Those luckless snails had spent sixty-four hours in the devil's own slow cooker. After two hours, they were escargot. By the time we arrived, they had long passed any state of decay that pathology has described.

Our foreman assembled us on the sidewalk to make a plan. One man was dispatched to a hardware store to buy every respirator they had. The foreman triaged the crew, sending the greenest half home. When the respirators arrived, the remainder of us went back upstairs to whitewash the damage. Every window was flung open. Gallons of bleach were dumped into the brook. We set up deodorizers, air purifiers, whatever might work. Any of us with a title above laborer was happy to volunteer ourselves in whatever way we could, if only to avoid the worst of tasks that we knew would fall to one man—Kevin.

No one deserved to do what he did that day. In pity, he was given the best of the respirators. Rugged gloves, boots, and coveralls were found for him. No amount of equipment could protect him from the horror of the thing, made worse when some manager asked him to save all the Colorado River rocks. Kevin spent the entire day, dripping in ooze, sorting one-and-a-half-inch smooth, round, brown rocks from one-and-a-half-inch smooth, round, brown snails. He scrubbed each rock and tossed it in the "clean" bin. He threw the putrefied snails into contractor bags and tied them tightly with twine.

The day progressed; the air cleared by measures. We took up our habitual tasks, occasionally looking Kevin's way as he moved the bags inside once they were filled. His movements had become jerky and agitated. His expression, once marble, had turned to schist.

We were startled by the foreman's sudden shout: "Everyone out! Now!"

There was no indication that anything had gone wrong. Not the usual smoke or screams—we were all working quietly. Even Kevin had finished up and was busy bagging his protective clothing for disposal. We huddled down the freight elevator to the street. There were several police cars and a fire truck pulled up at odd angles. The entire building had been evacuated and the residents were scurrying away from it into Park Avenue. The doorman was vomiting by the curb. And all around us, inescapable as death itself, that smell.

Modern buildings do their best to make themselves easy to live in. Heating and cooling are distributed automatically. We get water from the wall. Sewage is whisked away by stealth. Even garbage is gone in a breeze. This building was fitted with slick steel chutes, which led to a shipping-container-sized compactor behind the lobby, where garbage was compressed into landfill-saving packages, removed twice weekly by the city's trucks.

From the thirtieth floor, it is three hundred feet down that chute to the compactor's steel floor. I am not a physicist, but I imagine that by the time a twenty-pound bag reaches the bottom, it is going fast. Picture that same bag, or better, ten of them, filled with grapeshot-sized snails; snails so dead hell wouldn't have them. Now imagine those ten bags, traveling one hundred miles per hour, shattering and atomizing against the unyielding floor of the compactor, painting its entire interior with the dank reek of the rotting dead.

As most are, the job was eventually completed. A custom outdoor refrigeration system was built under the deck to counter the sun's cruelty. Walls were painted and repainted in indistinguishably better tints. Everything was made to function, at least until the next maintenance call. Kevin was never heard from again.

The owner, pleased with our work, threw a Christmas party for everyone involved, a rare gesture of genuine appreciation. He greeted us in his rotunda with the ambitious young architect who had designed the deck, now his fiancée, by his side. It was the first night out for my now wife and me after our son's birth. Once the group gathering for formalities and toasts had ended, people wandered off to look over the crazy quilt of finished rooms. I went upstairs to see the custom furniture; I almost never get to see a place once the owners have moved in. The night was cold and dark. A few couples ventured out onto the deck. I watched the men help their dates across the stepping-stones in the brook. As they made their way to the railing a few women wobbled, a knee forced straight, a gait undone, as a heel plunged into the soft soil beneath the barely visible stripes of grass.

CHAPTER 6

Attention and Intention

"Look both ways before you cross the street."
—Mom

The brain is an unruly place. Scientists estimate it contains around one hundred billion neurons; that's about the same number as there are stars in the Milky Way. Neurons are connected by synapses, bridging axons, and dendrites, all with the purpose of allowing parts of the body to communicate with one another. Who can tolerate living in such a cacophonous muddle of a switchboard?

Mercifully, many parts of this apparatus work automatically, largely without our awareness or participation. Complex networks tell our lungs to breathe, our hearts to beat, our kidneys to filter, and so on, all whether we notice their work or not. Other constellations of neurons push us this way or that, urging some action on our part without being specific about it: Get food, stay warm, find mate, save skin, that sort of thing. We have to figure out what to do, but our brains are insistent that we obey. A third class of circuits is employed in the activities we take up as we move through time. These are our habits and learned skills: I tie my shoes using the "bunny ears" method, always have. "Shall I make a nice stir-fry?" Some like these are pleasing and useful; others outlive their purpose, but still invite us for the occasional nostalgic rerun: "Dad was never the touchy-feely type." "All I wanted was for her to love me back." These we call psychological scars. They are our short circuits, favored routes of habit and repetition, free of discriminating road signs reading WRONG WAY; GO BACK.

A fourth class of wiring is left wide open for us to use as we wish. Undedicated to any specific purpose, these bundles wait for higher-amperage currents powered by combinations of physical, emotional, and intellectual circuits, and occasionally by all three at once. When these circuits are activated, we call it *meaning*.

Suppose someone lives in such an environment, badgered by the need to find the next meal and plagued by suppurated assurances that the future will be rife with failure. One day, they get the idea they want to do something compelling. The first thing to know is that any steps in a new direction are going to cause disruption among the routes and habits the brain has already established. Fatigue will set in, and the maintenance department will shut the system down for sleep; hunger will arise and it's time for a feeding; habit will find a way of reasserting itself every day; and there is always some old recording somewhere of a sibling or an instructor repeating endlessly what a fool you are to try such a thing. All of this is distracting and out of our control.

I come from a long line of men whom circumstance brought to maturity between wars. We like to think ourselves peaceful and understanding men, but I imagine that if the calendar had been shifted twenty years in any direction, my forebears would have been called up, trained for service, sent off to horrors, and we would all have turned out very differently. I'm grateful luck spared us. But it's interesting to think of the men in my family and wonder how completely unlike ourselves we would be if we had lived our lives in the military. I suspect that morals and beliefs that we see as guiding stars would be poles apart from where they wound up in us pacifists.

People are fond of saying how difficult it is to change. They're correct; biological necessity and habit conspire against it. But even as crude a system as the military knows how to rework a brain. It's the first word every recruit hears: "Attention!"

Gentler methods, at least until you try them, focus on precisely the same faculty. Tai chi, yoga, meditation, therapy, and all other systems of psychological change begin with attention. They do this because they recognize that attention is the *only* faculty we possess that we can reliably control. The first thing anyone who takes on one of these pursuits discovers is just how feeble the quality of their attention is.

Poor Attention, the untrained puppy of the psyche, sniffing at a stump one minute, then five steps later rolling in rot. But with patient training, the little scamp can be transformed into a hunting dog that chases down quarry, a guide accompanying its master down unfamiliar streets, or a rescuer that can find us buried in an avalanche and bring enough broth that we might work our way back to warmth.

I don't think one can overstate the importance of developing attention. It is rightly described as a form of payment. We invest our attention in the things we value, and we pay with our psychological energy. As with any organism or device, there is only so much energy at our disposal. If all of my attention is captured by the next fetching attraction or the next sad thought, then that is where I will spend my life. There won't be any energy left for anything else.

We live in an era that devalues and even attacks attention. Multitasking, constant distraction, advertising, titillation, gossip, rumor, undirected desire, worry, and most forms of entertainment steal our attention without us even noticing. So that is where we spend most of our psychological energy. I would never suggest that it is harmful, after a hard week's work, to enjoy a movie or a concert, but I would posit, in all seriousness, that half of our country's infants would not be mistaken if they were to grow up thinking that their parents loved their phones more than them.

Emotions, so unreliable in their assessments, grip me without warning. Thoughts stream through me unbidden and suspect in their origins. Sensations signal this or that function needs care. Attention is often the only capacity I have that can correct my course and lead me in a new direction.

In the winter of 1993, not long after my first son was born, I was hired by a millwork outfit to install the entire second floor of a venture capitalist's Connecticut estate. He was a kindly venture capitalist, polite and well spoken; he made a point of giving a tour of his lovingly tended rose garden to my crew. He may have behaved quite differently on the opposing end of a conference table, but up close he was affable enough that I had no objection to providing him with an imposing oaken office and a suitably masculine dressing suite.

At the top of his sweeping staircase, a high-ceilinged, faux-rusticated-stone rotunda fashioned from ornamental plaster connected with a private sitting room that opened at three compass points to wings of the second floor. To the east, an archway led to the master bedroom suite. To the south, an interior sunroom was sheltered from the elements by a wall of glazed French doors that accessed an expansive balcony overlooking the grounds. To the west, a pair of tall double doors opened into short passages leading through the bookshelves of the office on the wall's opposite side. The passageways tapered to wider openings on the office side. They were rich with raised panels and heavily molded. We installed both of these elaborate openings in a single day. They were mostly glued in place, visible fasteners being anathema to the architects of the well-to-do, so we had to wedge several two-by-fours across each of them at the floor to keep pressure on the glue while it dried overnight. This is what we pros call a "trip hazard."

Since we fancied ourselves responsible individuals, we found two cans of fluorescent spray paint, one flaming orange, the other flaming green, and went about painting garish tiger stripes over all of the offending two-by-fours. This still didn't seem like enough, so we made two poster-sized signs with the made-for-

a-billionaire slogan DANGER!! LOOK DOWN↓ We taped them across each opening at chest height and were satisfied.

We went about tidying things up for the evening and were about to leave when our client, who was well into his sixties, arrived through the dressing room door on the opposite side of the office, accompanied by that rarest of companions among the wealthy, his first wife. She was particularly taken with the coffered ceiling and asked if we had made the turned pendants at the apex of each run of beams ourselves. Her husband took a step or two back to admire them. Then another step or two back. Then another. Before anyone could say a word to stop him, he teetered. Watching a billionaire fall is a lot like plummeting off a cliff or down an elevator shaft. Having experienced all three, I can attest that the first is the most unsettling. There's a lot of time to think through the possibilities. I foresaw a future in prison stripes; I envisioned my only child holding out a tin cup, his mother just behind him, urging him to beg more pathetically.

Our client fell hard on his most undignified part, sprawling gracelessly. His wife turned to him, looked him over, scanned our sign and the Day-Glo two-by-fours, and pronounced, "Honey, it looks like they knew you were coming." Our client struggled to his feet, sheepishly dusting himself off, while I thanked heaven for the rare marriage that lasts longer than mine did.

WHAT WENT WRONG HERE?

The most beautiful job I ever built should have been counted as a success by everyone who participated, but it wasn't. There are only two jobs I have visited years after completing them, my favorites. The quality they share is that every task, down to the angle to which screws were tightened on hardware, was carried out in a considered way. To do this in a multimillion-dollar project requires an outright painful level of sustained attention.

I toured both with a writer who was interviewing me and the architects who had designed the jobs. Unlike the myriad articles that had showcased my work in the past without ever mentioning my name, the subject of this article was *me*. Burkhard Bilger from *The New Yorker* had shared one too many beers with a colleague of mine who frequents the same watering hole in their neighborhood. Burkhard got the intriguing but unlikely idea into his head that he could hold his readers' attention for fifteen pages by writing about the hidden world behind those glossy magazines' photo spreads and sought me out as his protagonist. At his request, I negotiated access to two of my most prized projects.

The second home had won awards and recognition; it was publicly hailed as a success, so everyone thought of it that way. The first was built for a couple who valued their privacy. Professional photographs of the apartment were taken for the architect's portfolio, but they weren't published, so the home never garnered the design community's attention or its blessing. The couple certainly didn't care; they adored the place. Perhaps the architect would have valued the publicity, but I know she is immensely satisfied with the results of her work. It should have been a success all around, and now, with time's passage, most of us see it that way, but the contractor I worked for back then would disagree.

Somewhere around the latest turn of a century, a majority of people started to think of digital technology as important. This sort of thing has happened over and over in history. I imagine our earliest ancestors exclaiming heatedly of an evening, "I'm telling you, this pottery stuff is really something! It's gonna be big! Bigger than spears! Bigger than hatchets! You mark my words!" And so on with bronze, iron, steam . . . Doubtless, computers have revolutionized the sciences and mathematics, fields that actually do computation. In construction, computers have supplanted conversation, leaving out all its context and coloration, and they have replaced file cabinets and drawing racks, robbing offices of their gravitas and musty charm.

Just such a construction office was the one the contractor on this job had built; every desk had a computer on it, and every field supervisor was shackled to a shiny laptop on which they were expected to tap out elaborate reports recounting each day's activities. Why, a contractor could know everything that happened in his corporate empire that day without ever leaving his X-Chair!

I was assigned to help a field supervisor with the layout and execution of this same three-story apartment in a splendid Beaux Arts townhouse near the Metropolitan Museum of Art. The supervisor was struggling to keep up with the innumerable details of the job. He was too busy with all those reports. I was spared the burden of a laptop, a boon, since I had no idea how to work one; all that was required of me was a monthly summary of our progress.

The architect was youngish and eager, but not so young that her schooling had neglected to instruct her in the traditional ways of her trade. She sketched well, in the detailed fluid style that was lost to the next generation. She was suited to the project; the clients were looking for a complementary setting for their extensive collection of Art Nouveau paintings and sculpture and she was eager to create it for them.

Everything behind the walls was meant to be up-to-date, but the decorative elements were from another era. She had inherited a thick set of drawings from the architect she replaced. The designs therein were classically ornate and bland, suited to the neighborhood but not to the needs of these specific clients. Letting the general layout remain, she set about reworking every visible element of the apartment. Staid Greek Revival doorways became gently arching passages with well-coiffed, deeply carved curves where they transitioned to the horizontal. Meander-bordered tile floors were garlanded with ribbons and bows. She sketched, modeled in clay, and conferred with us constantly, always trying to make the place better, more cohesive, more pleasing to absorb.

This is a terrible way to renovate an apartment. Renovation is akin to a one-way journey to a far-off place. The more one knows about one's destination, the easier it is to prepare provisions, vehicles, and personnel and therewith secure a pleasant passage and carefree stay. It's not too difficult to imagine the confusion created when one sets out on a trip to Classical Athens only to be told after embarking that the destination is in fact Belle Époque Paris.

Our schedule read two years; early on, it was clearly slipping away. The budget slipped away with it. The project manager dutifully cataloged every change, calculated its cost, and obtained the clients' approval for each of them. The clients appreciated that something special was coming their way, something only they would inhabit. The samples presented to them were sumptuous, tendrile, and unique. They could afford it—why not go beyond?

Heady as it was, from where I sat, there was cause for concern. My monthly reports grew grimmer as our deadline approached. Millwork that was supposed to have been in production six months before wasn't even designed yet. Half of what we were to build hadn't yet received the refining and reworking our architect intended. I sat at the super's computer, pecking at the keys, trying to formulate an unmistakable message to our boss. This is what I hit upon: "*We have abandoned all hope of meeting our schedule.*" Several pages of particulars followed. To this clear call for help, no response came.

So we carried on, completing the sweeping staircase with its hand-forged balustrade, a crackle-glass catwalk with cast tulip-bud stanchions, a leather-and-walnut-paneled gentleman's study, and so on in that vein. Nary a visitor walked through the job with their jaw in place; tradesmen, designers, prospective clients all agreed that magic was afoot.

Then our deadline came. The owners had let the lease expire on their temporary lodging and demanded a fixed move-in date. We were months away from completion and the boss hadn't said a word to them.

The husband was a man given to temper; my boss trembled at the idea of bringing him bad news, preferring to lean on his well-developed habit of obfuscation. But that man's day job was buying and turning around failing companies. Our project manager had presented more than three hundred change orders, amounting to millions of dollars in additional work. My boss had forbidden him from adding a single day to the schedule. The owner might not have smiled at the idea, but he could certainly have understood that 80 percent more work might excuse a 30 percent delay. A monthly update showing creeping cost and schedule overruns wouldn't have been joyously welcomed, but it would have given the owners a chance to understand and plan for the reality of what we were doing, and doing for their benefit. But out of fear, no one had been allowed to say a word. So a minor monthly upbraiding became a full-blown denunciation and scourging, and my boss deservedly bore the brunt of it.

The months that followed were less magical. Meetings were called; repercussions were felt. The project manager spent weeks trying to wheedle millwork out of the French company that excusably hadn't finished it yet. The relationship deteriorated, ending when the elaborately carved entry doors were held hostage like a count in the Bastille, only to be released when a quarter million dollars in ransom was paid. It was an ugly ending to a beautiful job.

The day before the owners moved in, my boss came to the site. He approached me directly, something he might have done with the owners nine months ago. "What went wrong here?" he asked. Nothing had gone wrong. The job went beautifully. Human interaction, or the sorry lack thereof, sold us all out. I went to the computer and pulled up my monthly reports. "What part of '*We have abandoned all hope of meeting our schedule*' was unclear?" He started in with a response. I wouldn't have it. I released a colorful twenty-minute torrent, unmatched by anything I've said to anyone before or since. When it was done, he left without a word. I supposed that I no longer had a job but I didn't care. The assistant project manager, Pauline, looked up from the desk where she'd been hiding. "That was scary."

She always knew the right thing to say.

INTENTION

> "What are you going to do with your life?"
> —DAD

In old movies, whenever an eager suitor comes around, a concerned father will ask, "What are your intentions toward my daughter, young man?" In the entire history of cinema, no one has ever come up with an honest reply.

Civilization has given us morals, manners, codes of conduct, laws, and punishments. They rein in our rougher impulses and allow people to interact with a welcome minimum of bloodshed. We have become so accustomed to our own civility that we hardly notice when, for instance, my barista says, "We're out of oat milk, would you mind soy?," and I answer, "Oh, sure, that's fine," while inside my head a voice rails, "You can't run a simple coffee shop, you fucking cretin? Tufts didn't teach you how to count containers once in a while?" I like to think that I'm nice, but apparently I'm not. It's laudable that nobody got hurt, and perhaps no one suspected a thing, but my shoulders hunched up, my face tightened, my voice changed in quality, and my barista could be forgiven for getting a faint inkling that I am not to be entirely trusted. I don't think I'm alone in living like this every day.

One advantage of growing older is the creeping realization that very little of what goes on outside or inside me is my doing. I didn't invent morals; I didn't invent outrage. I haven't invented much of anything, and yet noisome conflicts play themselves out inside me all day, every day. That's neither good nor bad; most of it isn't of my making. The trouble begins only once I decide that I want to do something.

Suppose I were to state, "I'm going to learn everything I can about my date tonight; I really want to know if we're suited to each other." That is an intention. Halfway out the door my boss calls to tell me, for fifteen unbearable minutes, that the client has lost their mind because their dressing room countertop is scratched. In all likelihood, the entire evening will consist of me holding forth on the difficulty of my work, the cosseted unreasonableness of my clients, and the exasperating inability of my co-workers to read an enormous sign saying NO TOOLS, FINISHED COUNTER!!!

That's the likely outcome, but not the necessary one. It is

possible to hold on to an intention, to repeat it, even out loud, and to ask, "Where am I in relation to my intention?" Not always, but from time to time, and more so with practice, it is possible to realize how far afield I have wandered, and to bend my path back toward the one I meant to be on.

SUMMER RULES

People who build for a living are often in the awkward position of knowingly making other people's lives miserable. We will come into your building and create months of dust, noise, elevator traffic, and damage. It's a wonder anyone will even let us through the service entrance. They wouldn't, except for the enormous industry that promotes the idea that the space a person lives in defines their worth. Glossy, overpriced magazines devote themselves entirely to this idea. Their pages are filled with three images on eternal repeat:

1. Stratospherically expensive city contemporary classic with sweeping views
2. Airy, quasi-rustic beach house
3. Country escape with surrounding gardens and at least one stone wall; Europe preferred

If you own one of these, you are on your way. If you own all three, you have arrived. Discouragingly, people then find that they have landed on the lowest rung of the next ladder, one requiring private islands, far more extravagant seaside estates, and groaning helicopter-bedecked yachts, leading inevitably to their distribution among coddled heirs incapable of managing the upkeep.

So why do fancy building owners let us through the door?

Because that ragtag collection of unsavory types waiting by the service entrance can give you number one from the list above. This is such a painful thing for some clients to admit that they hire design teams and owner's representatives to insulate them as completely as possible from the actual building process. In case I stand accused of exaggeration, I reference the dozen or so newly purchased toilets I have thrown into dumpsters because they were used by workers during renovations. There is no disinfectant powerful enough to sanitize that sullying.

Human awfulness goes by a lot of different names. When the purveyors are wealthy, we call it "snobbery" and "oppression." When the purveyors are poor, it bears more colorful names, like "the Purge" or "the Terror."

It is an unfortunate characteristic of our species that people are usually most comfortable when in the company of their own kind. Construction crews often do not look, speak, or act like the people for whom they work—maybe they do in Portland, but not on Park Avenue. I expect that to them we resemble criminals and gang members from movies in which well-to-do, educated folks are the heroes. As a result, our clients are afraid of us; they are uncomfortable having us in their homes. That fear manifests as suspicion and contempt.

I've spent long hours assuring a client that no one on my crew stole their delivery of Neiman Marcus sweaters, or even knows what those are. I have walked through a neighbor's apartment with photographs taken to document its condition prior to the start of work upstairs and been told that they didn't care if I had photographs proving that the cracks in the plaster crown were there before we arrived, we were going to pay to have the entire apartment repaired and repainted. I have been greeted at my jobsites' doors by the police, the fire department, teams of lawyers, and a retired federal judge, all informing me that if the complaints they were receiving didn't stop, they would see to it that my work did. The message has been repeated and received.

It is the rare customer who knows *any* of my crew members' names. Some can't even remember mine.

At the same time, to my crews, Park Avenue people look like the smarmy, scheming swindlers from movies in which working-class folks are the heroes. We gleefully ridicule them for their weak constitutions, dependence on psychoactive drugs, failed marriages, and spoiled, ineffectual spawn. I have had only two or three clients in my career whom I would consider suitable babysitters for my children.

For the most part, both camps seem quite content to maintain the chasm that class carves between them. It's a shame, and an opportunity lost. Primary among the elements that keep building interesting to me is the unimaginable mix of people who carry out the work. People I work with every day have lived through difficulties no newscast can convey. Those who come to the site regularly find remarkable ways to get along despite differences that might make them enemies anywhere else.

This has vastly expanded my idea of who is "my kind." It's no mistake that the few clients for whom I feel fondness are the ones who visited the site often, and who showed the most interest in us workers and the work we do. A lot can be gained when people get to know one another; a lot of kindness can be learned, but people conspire to keep attitudes as they are, and meanness is the result.

Around twenty years ago, city co-op boards came up with a clever plan to avoid encounters between homeowners and renovation teams. A co-op board, for those who don't know, is a collection of successful professionals in non-building-related spheres who, under the assumption that only underachievers are forced to wield tools for a living, decide as a group that governing a building is a breeze. The obvious solution to the "contractor problem" was to allow renovations in their building only when the habitants were sure to be out of town: from Memorial Day till Labor Day.

Since this solution's implementation, as affluent co-op owners break out their summer whites, workers break out their jackhammers and wreak three months of havoc on New York's poshest addresses. Then, just as the seersuckers are swapped for worsteds, contractors pack up their trucks for the year, no matter the condition of the work they are abandoning.

Some simple math reveals that, under this system, a fifteen-million-dollar renovation that could take two years if run continuously might now take eight. Even the co-op boards puzzled this one out. Again, their solution was plain, though varied by idiosyncrasy. In many buildings, no renovation may last more than four years. Some boards go further still, disallowing the use of pneumatic tools and rotary hammers, or insisting that all plumbing be rerun in threaded brass, perhaps in an effort to keep plumbers from racing about the place too freely. One can't blame contractors for awaking, soaked in sweat, from Kafkaesque dreams that they are cockroaches being forced to run a maze, with ever-increasing difficulty, after a white-suited doctor removes another leg at each starting gate. Yet, from the co-op board's point of view, these are reasonable requests, and even new owners' complaints cannot change their minds. Co-op boards are made up of people whose apartments have already been renovated.

That point brings me to a couple I met thirteen years ago. They were a handsome pair. He was deferential but sharp, comfortable enough to stand on debris in thousand-dollar shoes, his reticence not quite covering a contract lawyer's flair for methodically removing the floor from beneath one's feet. He was an achiever. His wife was straight out of the Bouvier mold, from low-heeled Arpels to low-rise bouffant. Her observations were polished, her comportment finished; she had an expert's eye tempered by a considerate, decorous path to a point well made. She was a prize. We in the trades are so unused to appreciation that when we encounter it, we will redouble our efforts to do right in return. I would warn prospective clients against the contrapositive of this proposition.

The entire contract, for a three-thousand-square-foot renovation on East Seventy-second Street, came to about $5.3 million. What would usually be a ten-month schedule was reduced to three. These numbers aren't meant to shock with their magnitude. The first time I worked on one, I said something like "Can you believe it, a MILLION-DOLLAR PROJECT!" Now we build a kitchen or a staircase for as much. Instead, I mean to show that we were charged with building this apartment three times faster than the already demanding standards that New York clients require. Ten years hence, the amounts will seem laughable, but the challenge will remain the same.

The building's rules were straightforward:

1. No one but managers were allowed in the apartment before Memorial Day.
2. Noisy work could progress until Labor Day.
3. Painting and decorating could continue for two months into the fall.

In order to meet this deadline, a small army was assembled. Five millwork companies would build the cabinetry. Two flooring companies, two plaster teams, three marble fabricators, and redundantly on in the hope that everything could be measured, made, and installed in time. I went in early with an assistant to lay out, sketch, and preorder anything I could. Shop drawings were submitted, corrected, and approved for fabrication before there were even walls to attach things to.

The day after Memorial Day, I welcomed a demolition crew of twenty at the door. In one week, they tore the place apart, cleaned it thoroughly and removed every trace of the previous décor, leaving a scraggled, dusty shell in its stead. The next week, just as many carpenters arrived with tools, subflooring, and studs. We were off to the races. My assistant and I snapped chalk lines on the floors and ceilings while hard on our heels, carpenters followed erecting walls, door openings, closets, and niches for which the wood and stonework were already being made. Then came the hidden trades: plumbers, electricians, and HVAC contractors. Their equipment had all been ordered in advance. Loads of it were dragged up the service elevator nearly every day.

Life is full of little fiefdoms. DMV workers, tax clerks, traffic cops, and border agents might have limited agency outside of their bailiwicks, but within their designated spheres, they rule supreme. Service elevator operators are a sleepy breed. Perhaps neurologists will someday discover that brains designed for wandering thither and yon are adversely affected by perpendicular travel. From behind their pantograph gates, they are the sole arbiters of who goes up and who goes down. One may dislike them, find them coarse, even capricious and unfair, but they must be befriended. This is a useful exercise. With few exceptions, everyone has redeeming qualities, people who count on them, and hidden capacities and histories. Uncover one of these, inquire about it, place several hundred-dollar bills in a plain white envelope, and make a new ally.

Summer moved quickly along; one could almost hear the clock ticking through the calendar days. On occasion, the husband would return to the city for business. He would stop in, express admiration at our progress, and repeat the same question at each parting: "Are we going to make it?" No matter how much I wanted to say yes, I couldn't and wouldn't. It was too much to promise; the final decision was in the hands of the building management, and there were too many things that might go wrong.

Of all the jobs I've managed, this one had the fewest mishaps. We had a cooperative and expeditiously inclined architect who understood that, as our schedule demanded, I was going to make on-the-spot decisions without him. We had already done a project together, and he trusted my judgment. We had subcontractors who manned the site at whatever levels I asked, all of whom worked unflaggingly toward our goal. Even the building's administrators seemed to be on our side. The co-op board president and building's resident manager would stop in from time to time, tour the place, and leave with an encouraging word. Anyone who has ever had a collaborative job knows that it is a rare thing to work in an environment where so many people's intentions align. But it was still too close to call.

As the final two weeks approached, the situation was grim. Everything was nearing completion, but every trade had a substantial way to go. I drew up a list in the minutest detail, cataloging every remaining step. It was more than one thousand items long. There were ten workdays left. If we could assemble a crew of fifty, each worker would have to complete two tasks per day.

I spent the rest of the afternoon writing names next to each task and called my boss for reinforcements.

On Monday, two of my supervisor colleagues arrived with my boss to help. They brought ten additional workers from their projects, as many as they could spare. With the subcontractors' crews, we were up to fifty-four. We worked through the week and completed 40 percent of the tasks on my list.

The next Monday, the boss arrived ready to work in a T-shirt and jeans; he brought a five-person laborer crew with him whose only job was to keep order. Five more carpenters arrived shortly after them. The subcontractors had all been asked to boost manpower for one week; they delivered on the request. For four days, seventy-two workers, their tools, materials, and equipment inhabited a three-thousand-square-foot apartment in one-hundred-degree heat, working shoulder to shoulder toward our common goal. My boss's willingness to work along-side them did wonders for morale and fellow feeling, despite the fact that each person had a workspace little bigger than a single sheet of plywood. I had one job: complete the list. If anything new came to light, it was added; as items were verified complete, they were crossed off. If someone struggled with a task, I reassigned them to a simpler one and finished theirs with a more confident worker.

Thursday evening came, and the board president and resident manager appeared at the front door. Unless they were veterans, I doubt either had witnessed such mayhem. Sweat-soaked, dust-streaked workers were everywhere, thick as a bramble. Plaster dust filled the air and coated every surface. Laborers scurried about with trash cans clearing debris before it hit the floor. Stacks of every material were placed in the center of each room so tradespeople could work in the remaining space near the walls. The men asked for my boss and me, unable to distinguish us from our troops. We got word that we were summoned and approached the two men from opposite sides of the foyer. "What can we do for you?" my boss asked.

"We're here for the final inspection. Fellas, it looks like we have to shut you down. This place is nowhere near done." I looked at my list; there were forty-three things left to do. I turned it so that the two men couldn't see and said, "You're early." "Yeah, but seriously—" he started to reply, but I cut him off. "We have twenty-four more hours to get this place done; Labor Day is the deadline." It was a bit of hyperbole; the building allowed only seven and a half hours of construction per day, but they both heard my point. "Come back tomorrow at four. That's the agreed-upon time." Both of them shrugged, assented, and left. They were openly dubious, but fair.

My boss and I turned to the list. We had enough workers to assign each task to an individual. We called a quick group huddle, gave everyone their job for the next day, and asked if they were up to it. Two or three people doubted whether they could complete their assignment; they were each given a partner or traded with someone for an easier task. Everyone else, the boss included, was designated a cleaner. Two laborers ran to the local hardware store for cleaning supplies, and we all went home for the night.

The last day was dreamlike. The weather gods sent us a welcome eighty-degree day. Two laborers spent the entire morning taking orders—water, coffee, sandwiches, whatever anyone wanted. My boss paid for all of it. Forty-three tradespeople worked at individual tasks, calling me over as they were completed. As each task wrapped up, that worker took their tools down the freight elevator and went home. Twenty-five cleaners removed anything that wasn't about to get nailed to a wall. I monitored the list; my boss swept. As the day drifted to a close, the jobsite cleared by degrees. It was all weirdly simple and orderly.

Finally, four o'clock came, and the doorbell rang. My boss and I answered. We were covered in grime, but in the apartment, only ten workers remained. They were packing their things to go. The floors were clean, the materials and debris gone. Each room was complete: Sinks and lights worked; doors and molding were installed; counters were empty and shining. Neither my boss nor I had taken a moment to really look at the place. It needed paint and wallpaper, but it was built. The co-op president and building manager walked through each room with us, testing every faucet and switch. Nobody said much, just an occasional "Okay" or "Good, that works," as we wandered through. We made our way around to the front door. The manager turned to the co-op president and asked, "Pass?"

"Yeah, pass," came the answer.

The president turned to us, and looking back and forth between us said, "Okay, you pass. Nice job." The two men shook our hands and left. My boss swung his head from an exhausted slouch; his face was smeared and slack, with the hint of a bemused smirk. He lifted his hand to my shoulder and said, "Could you call the painters for me, please? I want to get home." "Sure, I'll set them up for Tuesday. I'm heading home, too."

"Have a good Labor Day. Get some rest," he said.

"Yeah, you too."

We shook hands and left.

Every day, I go to work and play an elaborate game. It's a game with hundreds of rules, some dictated by building codes trying to remedy the dangers of a lax past, some created by contractors' cowardly attempts to meet their clients' unrealistic expectations, some inspired to preserve the city's history even when it has long been lost to changing tastes, and some by building boards who wish we could be banned from their basements. It's a game that no one is expected to win outright. Most don't, but every once in a while the stars align, and the impossible happens.

CHAPTER 7

Competence

"Are you any good, kid?"
—*A LOVABLE SKEPTIC*

It took about twenty years for me to feel that I had reached the level of competence in my trade. For those given to calculation, that's about forty thousand hours of work, which puts me in the slow-learner category. It has taken another twenty years for me to feel competent at what I am doing now, which embraces several professions, trades, and collaborators, as well as the financial and client relations side of construction.

Twenty years ago, I was willing to call myself competent in carpentry once I had developed a broad understanding of the materials, methodologies, principles, and limitations of my trade. If someone came to me wanting to make a secret pivoting door with disparate finishes that blended into the walls on both sides, I could say with confidence that I could build that. On the other hand, if they came looking for a carbon fiber and stainless steel rotating balcony with built-in fiber-optic lighting, I would have to say that they had the wrong guy. Twenty years hence, I can say, "Sure, we can do that."

Competence at anything is a charmed, relaxed, and temporary station. I have felt it in a few pursuits; I am there now with my guitar playing after fifty-five years of practicing music, forty-seven of them on guitar. Finally, when I play, I can say, "Oh, I liked that!" or "That was an inventive little passage!" The satisfaction is accompanied by the real pride of accomplishment. There is an ease to the endeavor, whereas before there was always noticeable effort.

And yet, unease finds its way into everything I continue to pursue. At forty, my purpose was clear: pay the bills, raise the kids, create a life that was as pleasant, productive, and satisfying as I could manage.

Now I'm almost sixty: The kids are on their own; the marriage is over; my career is on as firm a footing as it has ever been. I find myself with a head full of ideas, things I have never done that might put everything I have learned to good use. It is time to give up on the comfort of competence and move on.

But before I do, I will linger pleasantly for a while on a few things competence has taught me.

TEACHERS

"Your motorized mouth makes you monotonous."
—MRS. CARLIN, EIGHTH-GRADE FRENCH

The world is upside down. People who can repeat words someone else wrote while looking sexy or broadly likable get paid enormous amounts of money and receive inordinate levels of attention. Other people get paid to ask them their opinions, make them cry on camera, or share a petty secret. Another horde waits raptly for the next scrap of information they release, or to see oversized-them battle a screeching foe drawn in later by cloistered computer whizzes. Meanwhile, back in our schools, underpaid teachers reach into their own pockets to buy a few erasers for their classrooms' blackboards, in the hopeful knowledge that mistakes can be reconsidered and fixed, if only they can hold the students' attention long enough.

This begs the question: How do mistakes get fixed?

Mistakes get fixed by those who have the understanding needed to examine a problem's component causes, weigh possible solutions, and implement practical processes to rectify them. In my professional world, those three steps can be carried out by a single experienced person. That is essentially my job description. Blueprints appear on my desk. I take them to the address in question, and I spend the next few years uncovering endless errors that I will never build. Something that looks just like those mistakes—but works—takes their place.

We live in a time when the most pressing priority is to fix the errors of our past. We no longer have the luxury of imagining that everything is fine, the professionals are on the case, and there is plenty of time to solve our most urgent problems. These problems are thorny and too complex for one person to crack. Designers and engineers must find ways of communicating with technicians and users in respectful recognition of each group's expertise and experience. This is where most of the trouble develops in my industry, and it is where the most attention is needed if we are to avoid a whole new crop of calamities.

Halting climate change means reworking entire industries: energy production and delivery, construction, agriculture, transportation, regulation, political discourse, and more. But none of these areas will be significantly changed unless we upend our priorities. If few people participate in solutions and others actively work to thwart them, what will be solved?

Education is everything. I don't mean just the formal 8:00 A.M. to 3:00 P.M. daycare centers and children's prisons that masquerade as schools; I mean education in all its forms from parenting to professional mentoring, friendships, fieldwork for designers, theoretical instruction for technicians, in all possible fields. Teachers are everywhere, but there are not enough good ones to go around, and they are criminally undervalued.

We are making heroes of the wrong people, and at a time when we actually need saving.

I was recently reading about the development of the F-35 Joint Strike Fighter program. If you're reading this a hundred years from now, that is the laughably slow, clunkily shaped flying computer glitch that the Defense Department had poured more than $1.5 trillion into as of 2021. Even in your day, that's probably a lot of money. In our time, the cost is unimportant because it is critical that we kill more efficiently than our adversaries, preferably in a way that is automatic and free from conscience's interference.

All sorts of things have gone wrong with the F-35: At first it was too heavy; it lost its "invisibility cloak" when it flew fast; it malfunctioned when it was too hot or too cold; it became uncontrollable after evasive maneuvers. So teams of engineers, most of whom had never flown a plane, identified, analyzed, and designed solutions to each problem as it occurred. Now a small fleet of workable planes has been assembled for home use, and the rest are available in showrooms for friendly customers worldwide. It's great that our country has produced enough creative, clever thinkers to look after our well-being in this way. Perhaps the preservation of life merits the same level of national investment as its destruction.

One and a half trillion dollars is coincidentally enough money to send every high school senior in the country to MIT for four years. Now, MIT wouldn't take them: It's highly selective, and most seniors wouldn't go anyway, having lost interest in education way back in grade school when their teachers couldn't even come up with serviceable blackboard erasers. But it's interesting to imagine what might occur if we lived in a country that put as much value on educating its populace as it did on eliminating the perceived threat that foreign populations represent.

I have had the good fortune of receiving much of my post–high school education at little cost; often I've been paid while obtaining it. In school I was bored; I had difficulty seeing the point of the exercises I was assigned. In life and work, I have been an eager student; learning comes easily now, and I find teachers all around. In my work, there is always a new technique to learn, innovative material to understand, or a more exact or efficient way of doing something; usually there is someone around who is happy to demonstrate. Nearly every day, I learn something that had never occurred to me before. Likewise, nearly every day, I practice a technique I learned long ago, and I am often able to remember who taught it to me and when. Appreciation is not something that came naturally to me, but this is the way I've learned it.

People enjoy teaching someone who wants to learn. School is enervating to many because different children want to learn different things, and in different ways. Many students become frustrated with an education that doesn't speak to them, and teachers are equally frustrated that they don't have the time and flexibility to reach the students whom standardized education leaves behind. But cultivate an eager student, find a willing teacher, and provide the circumstances in which they can interact freely, and the world will be that much better tomorrow.

There is no single path. Why should there be? Some people learn well in classrooms, love new facts and ideas, read well, study attentively, and progress into fulfilling careers and lives. Others can barely get through a book but love processes, touching things, testing them, manipulating tools and instruments in ways that satisfy the learner and produce a useful result. There are at least a handful of ways that people can find their path to satisfaction and productivity. Our standardized system of education abandons far too many, forcing them to take their own road, where help is harder to find.

The relationship between student and teacher has changed markedly through time. Historically, apprenticeships were often forms of indenture, so their abolition was seen as liberating. These liberated learners were rounded up and corralled into classrooms where uniformity rules their days. I had antsy difficulty with this arrangement as my education wore on. I'd read enough history to know I didn't want to sleep on a shop floor and endure the occasional thrashing, but I would have welcomed the opportunity to learn a trade at someone's side. Any opportunity to get up from my chair would have been a relief.

Apprenticeships need not be usurious, and they might be formalized enough to be recognized without being robbed of their efficacy. Our world admires certificates, especially of the name-brand variety, ignoring the fact that most diplomas assure nary a whit of practical experience. The informal world of one-to-one learning subverts this system. Diplomas mean something in fields that require formal credentials, but they are no gauge of competence, skill, or care. In many pursuits, diplomas mean nothing at all; no one has ever asked to see mine; all anyone cares about is that I can do the job reliably, and with enough manners not to scuttle the company ship.

LEARNING

*The pile of things I know is always smaller than
the pile of things I don't know.*

People have been making things for a long time. These days, we zip around from place to place in all manner of transportation wonders, awaking in Scarsdale and nodding off in Kuala Lumpur. One hundred fifty years ago, most people lived their entire life within fifty miles of their birthplace. Perhaps many suffered from wanderlust's pangs, but if they did, their urges were likely well suppressed by the sunny desire to avoid drowning off Newfoundland in a gale, or to retain enough scalp to make use of Grandma's ivory comb. Most people stayed put; they farmed or learned their crafts and occupations from local elders who had learned them in the same way. The materials used were mainly those at hand: stone from the ridge, gravel from the riverbed, timber from the forest, wool from the sheep. They were combined and worked in simple ways, with tools refined through millennia, yielding an unimaginable variety of styles, forms, and methods that make up the endlessly inventive array of traditional local cultures. Any individual craftsperson might contribute one or two simple variations or improvements to a traditional way of working. The occasional standout might take a great leap forward; a local lunkhead might set things back a few generations. But the deliberate drive to refine and improve has brought us museums full of wondrously clever artifacts; the accumulated intelligence of an entire lineage of craftspeople is on display in every object therein.

Since those quaint days, railroad barons, firearms manufacturers, resource-hungry imperialists, and a host of their compadres have seen to it that much of the wilder world was tamed. Whizkid university students and political revolutionaries have dreamed up a future of never-ending harmony, leisure, and convenience. "These hidebound brutes are desperately trapped in their outdated ways. We will free them from their incantations and tired traditions! Enlightenment is what's needed here! Bring the broadest broom you can find so we can clear away these cobwebs!" said any of the above.

And so entire traditions have been lost—embroidery, thatching, beading, rug tying, chip carving, wood turning—consigned to the dustbin of decorative arts, replaced by gigatons of concrete, steel, and asphalt. Door frames have been stripped of their casings, with never a thought given to the ingenious way they form a box beam at the jamb, capable of withstanding the most aggressive slam, all while dressing the opening in any style one might use to prettify this too-drab world. Away went the plaster crowns that softened the crease between horizontal and vertical so a room envelops its dwellers rather than imprisons them. Craft has been relegated to roadside fairs featuring clunky garage-made handiwork, when its greatest examples once graced palaces and cathedrals.

I don't quibble with modernity; I love my little pickup truck, efficient in its day, and versatile in its uses. And I can't wait to buy a brand-new electric replacement, although I have no need of a touch screen to adjust my mirrors. The question is, are we innovating intelligently in order to solve urgent, looming problems, or do we crave novelty for the attention it brings, its mantle of stylishness, and its illusion of progress?

Enamored of the new, we have lost the intelligence of the traditional. For millennia people have lived in deserts and kept their houses cool. Closer to Earth's poles, people built massive masonry stoves, some with built-in bunks for a cozy winter's sleep. We have inhabited our landscape for a long, long time. Most of us have had electricity or gas for less than one hundred years; some still don't. If we are to solve the problems of our time, we must look to the past and the future with equal vigor.

THE HUMAN CHEESE GRATER

Two years ago, I got a call from a friend asking if I would help with a staircase one might call "architecturally ambitious." I'm customarily curious about the things people dream up to build, so I agreed to a meeting. We gathered at the appointed time and assembled around a computer screen to stare at a succession of images. The photos showed artworks in which a wide variety of free-form elements were constructed out of what looked like carefully seamed fishing nets. They resembled the wireframe models computer drafters have been working at for many years now; they also reminded me of the string art I used to make as a kid with lines of nails and my mother's colored thread. Finally, the slideshow stopped on the lone image of a ghostly suspended stair. Here was the inspiration for the task at hand: "The centerpiece of the project will be a diaphanous staircase leading from the foyer to a skylight-covered bulkhead sixty feet above."

One gets used to this sort of sentence. In the initial stages of contact with people who speak like this, there is no use pointing out that carrying an air-conditioning unit up five flights of fishing net is likely something that OSHA would frown on. I can rightly be accused of stodgy traditionalism. To me the words "diaphanous" and "staircase" go together like "gourmet" and "oatmeal," which is to say, not at all.

Staircases are one of the few remaining architectural elements that inspire inventiveness and pizzazz. Inventiveness can have its shortcomings. Sometimes the most obvious failings are missed by innovators. There has been a spate of glass-treaded stairs going up around my city. They work hard to add back with modern élan what they subtract from many people's excusable expectation of modesty. But pizzazz is my market; it is the last bastion of challenge and fun in this industry. My attention was captured.

Patience is its own reward in instances like this. It is also its own penance. Drawings were ceremoniously spread about. It was revealed that the treads and risers of the stairs would be made from a continuous zigzag bent ribbon of perforated steel. Perforated just as one might imagine, with little quarter-inch holes as densely packed together as steel would allow and still remain a viable sheet of steel. A sample was fetched from a nearby shelf, about six inches square and one-eighth inch thick, already bent once to show the ease of doing so. I spent the rest of the meeting flexing it between my thumbs.

There is an enormous difference in graciousness between dashing someone's dream and refining it to fruition. I could picture the staircase stretching toward the skylight, countless conical rays spilling through it, creating unpredictable patterns over its sprawling height. The design wasn't something I would have conceived; I'm not even sure these architects had dreamed it up; the idea may have been cribbed from a recent magazine spread or been encountered by chance in a SoHo store. No matter, the newly constructed image of it in my head was engaging enough that I wanted to see it realized. But, flex, flex, flex, this stuff wasn't going to do the trick. An actuarial sense of duty compelled me to say so. Somehow, the mistaken idea that this was the thickest version of this material available found its way into the discussion. At the design team's insistence that this was their material of choice, I agreed to make a full-size mock-up of a few stairs and put it to the test. There was never any doubt in my mind what the result would be, but my mind had no allies at that conference table.

I get paid handsomely to do this sort of thing. No one has ever paid me anything to say, "Based on my long professional experience, I can assure you that this won't work." But architects are perfectly happy to have their clients shell out tens of thousands of dollars solely for me to prove that an assembly I know will fail will fail. I have grown in my ability to suppress feelings of guilt in these dealings, unless I grow fond of the client, in which case guilt floats right back to the surface.

I made the mock-up. It failed in two ways:

1. With my bathroom scale as a gauge, and my truck's pump jack as a piston, I applied four hundred pounds of pressure to the center of one stair. Where the pressure was applied, the stair gave way three-eighths of an inch. That is a lot. Most people would feel excusably queasy about descending a see-through staircase with a pronounced bounce.

2. Across the entire length of the bent noses and heels of the perforated stairs, the steel tore the tiny connections between the holes, not enough to separate, but enough to

make a serrated line of barbs along each bend. A staircase made of bread knives might command some interest in an art gallery, but it seems out of place as the centerpiece of a private home.

I took scads of pictures to make it clear that things had gone as expected: badly. I checked with a stair engineer friend who responded informally that quarter-inch-thick material was the minimum he would entertain for this purpose. I dashed off a report containing his suggestion and sent it to the design team. In response, they sent internet clippings of a small enclosed attic stair somewhere in England that had been successfully fabricated from precisely the material they had proposed, and which I had just tested to failure. I pointed out that the depicted stair was two-thirds the width of the one we were hoping to build; said staircase was entirely walled in and turned twice so that very little could be carried up its cramped treads. On top of that, it only ascended ten feet, so even if it collapsed completely, the damage would be limited. The architects ignored my engineer's recommendation entirely, dismissing the thicker material as an obvious abomination that would destroy the very diaphanous-ness they were hoping to achieve. Five years at Yale had left their self-assured mark on this pair's thought processes. Learning gotten from books brings nothing of the understanding that hard-won failure emblazons in us. Coaxing this concept to viability was proving harder than I had anticipated.

Sometimes colorful language can paint a more persuasive picture than a reasoned argument can. The uppermost story of this home was intended as an art studio. The life of a building can be long, I began, especially one that is designed and engineered intelligently. I asked the team to imagine some day in the future when a new owner moves into the townhouse. The spacious art studio will work perfectly as a music room! A crew of six stout movers is assembled; a Steinway is dropped at the doorstep, and the rough job of getting it to the top floor is undertaken. The group grunts its way through the first three levels, but the top flight is the longest and no one is looking forward to its challenge. They draw deep breaths and brace for the final stretch. The combined weight of the piano and crew tops two thousand pounds. Through the previous ten years, those stairs have seen a lot of flexing, what with holiday parties on the roof and rambunctious teenagers bounding about.

Engineers obsess over terms like "ductile failure"; architects generally read them once or twice in school and forget them. Our current scenario would make a wonderful breed standard for the concept. Our dedicated crew works its way to the midpoint of the stairway's longest and highest span. The history of daily flexing, having created a lacework of delicate fractures, allows one last groaning bend. A springing sound rebounds off the glass of the skylight above, followed by a creak, a snap, and a low long splinter. Several of the stair noses tear apart at their junctures, turning their bared jagged teeth toward the startled workers. They stumble and fall against the pitiless steel edges. The first few drops of blood spatter the white tiled floor far below, like an assault team's lasers before the onslaught. The staircase separates into halves, no longer supporting the hapless climbers who have nowhere to go but sixty feet straight down after suffering the flesh-tearing edges of each failed bend. The Steinway gets hung up in the handrail; it hovers there for a moment until the thin oak band gives way under its weight. The tiled floor feels the impact of each worker as he strikes; the Steinway is the last to make its brutal way to the bottom.

"I don't know about you, but I don't really want that on my conscience."

The staircase is now complete. It is something to see, although the first few times I walked down it I had the queasy feeling that I am more substantial than it is. It's an illusion; you could hang a pickup truck from any point on it if that were your notion of art.

It never occurred to me at the beginning of this job that I would eventually be hired by the owner to bring the same sort of scrutiny and technical revision to the entire project that I brought to that staircase. The process was infuriating. Not a single aspect of the home was conceived for constructability.

I have benefited by my history, which includes a healthy dose of failure. If asked to build a spaceship, I wouldn't have any problem saying, "I have no idea how to do that." I don't find the same level of self-acceptance from those who hold precious, over-priced degrees. Some people succeed marvelously right through graduation and well into their careers without ever doing the thing they profess to instruct others in doing. I don't think the designers of this staircase have ever built as much as a birdhouse with their own hands. Yet, at no time in the process did I ever hear them say, "I don't know how to do this," or anything near it. I can't imagine a more effective roadblock to learning anything new.

KNOWING AND UNDERSTANDING

Everything is my fault.

Things go wrong for all sorts of reasons, and when they do, suffering almost always ensues. Unlike some things in life that seem law bound and orderly, mistakes and their consequences don't appear to have any ratio by which they relate. On a project I did near Gramercy Park twenty-odd years ago, I was installing a steel fireplace mantel near a series of one-hundred-year-old leaded windows. I was cutting the steel with a handheld grinder and didn't notice that the sparks were spraying onto the glass behind me. Another carpenter stopped me. When we inspected the windows, you could see tiny pits where the molten sparks had melted into the surface of the glass. Before that day, I didn't know grinding sparks would do that. I found some Windex and a soft rag and cleaned the surface as best I could. Little black spark centers were still visible where they had melted below the glass's surface, but on windows that old, it just looked like character. We didn't worry about it much and no one ever noticed. I still feel a tinge of remorse about that mistake. That tinge is called *knowledge*.

Five years later, in a Central Park West penthouse, a metalworker on my job did the exact same thing when he was cutting away an old steel clip near a sitting room window. The same little molten sparks melted their way into a picture window eight feet tall by twelve feet wide that had one of the avenue's best views. We rubbed the window for half a day and couldn't make the little black spots go away. Then I called a company that had special glass-polishing equipment with a clever suction-cupped window-grabbing-and-grinding mechanism. They polished it for a day, removed the spark centers, and left a two-foot-square hazy area that even the most medicated of my clients couldn't have missed. Two weeks, one crane, and twenty-five thousand dollars later a new window had been hoisted up the side of the building to replace the ruined one. The ten-second lapse in attention was rectified. It was the same mistake I had made years before, but with exponentially costlier consequences.

There's a lot of psychological ground between "a tinge of remorse" and "Two weeks, one crane, and . . ." Fifteen years on from the second mishap, neither I nor anyone who works with me has made the grinder mistake again. On my most recent project, I issued at least twenty-five reminders urging workmen to exercise the utmost care when using their grinders around glass and laborers to keep windows covered with film *and* protective plastic. That "tinge" was not enough for me to properly learn my grinder/glass lesson. The yelling, embarrassment, shame, and outsize penalty of the second incident hammered it home soundly.

This is the lesson that no university can teach. It is called "understanding."

FLOOD, FUR, FIRE, AND FISH

Eighteen years ago, I got called to a job on Park Avenue when I worked as a carpenter for my first truly boutique "high-end" outfit. By the time I arrived with my tools, the place had already been gutted. There wasn't anyone around and no one had told me what I was supposed to do, so I sat down and leafed through the blueprints that were spread out on a worktable.

On that day and for the first week, a "supervisor" would show up for twenty minutes around lunch, tell me he was about to start running the project, and disappear. It was mostly just me and the blueprints. Lacking anything else to keep me occupied, I examined them closely from front to back. They were unimpressive and riddled with errors—doors were designed to self-destruct, lights crashed into air-conditioning ducts above the ceiling, the usual sort of trouble. My perusals made the first day pass quickly.

On the second day I found a red pen and went through them all over again, writing questions wherever I found shortcomings. On the third day, I found a notebook and wrote all the questions into an outline, organized by the trade to which each question pertained. By day four, I was out of questions, so I reorganized my outline and rewrote it adding my suggested answers. Because answering questions is harder than asking them, this took me through the end of the fifth day. There was a fax machine on the worktable; I figured out how to use it from the instruction booklet and sent my outline to the company's office, Attention: Boss. I had worked slow weeks before, but I allowed that this was the first time I had worked an entire week and done absolutely nothing.

When I came back to work on Monday, the company's owner was onsite. He was a modern boss, given to referring to the people who worked for him as a "community." Occasionally there would be "team building" exercises and reviews that had the feeling of therapy sessions. It may have worked; a number of lasting friendships came out of that place, despite the sarcastic carping that went on whenever he turned his back.

Things had changed over the weekend. My "supervisor" had been fired and I had been assigned to replace him. I had run a few projects before, but nothing of this complexity. Modern Boss pulled my outline out of his briefcase and expressed that he had never seen anything like it. His chief sycophant, who accompanied him, praised it as well, naming it "The Codex." (A touch of flattery feels nice at first, but things between Chief Sycophant and me went to pieces in a year, so the term has since lost its charm.) A meeting had been arranged for that afternoon.

Modern Boss, the architect, Chief Sycophant, and I were all to attend. I now had a position of responsibility. In a single weekend, I went from having nothing to do to having everything to do. Nothing was mentioned about a concomitant increase in pay.

Modern Boss had a few traditional leanings:

1. He hired qualified subcontractors: plumbers, electricians, air-conditioning installers, etc.
2. He listened to people who knew their realm better than he did.
3. He believed in developing good crews from within; one could start as a laborer and grow into a supervisor.

Chief Sycophant's primary interest was pocket lining. To this end, he advanced the following:

1. The Smallest Tool Theory: The idea is to hire the weakest, cheapest subcontractors one can find. Using personal greatness, a supervisor was to organize, browbeat, and then threaten subcontractors into doing their work to a premium standard. (To test this theory, ask a woodworker or metalworker friend if it's fun to bore a half-inch hole with a four-hundred-pound cast-iron drill press. Answer: It sure is! It's so smooth! Now ask if it's fun to bore the same hole with a hand-powered push drill.)
2. Don't listen to anyone "beneath" you in rank. All listening efforts should be directed at Modern Boss so that his favor might be curried.
3. Use promises of advancement to garner workers' loyalty. It helps to hint now and then that so-and-so is "not going to last here."

At our meeting, they determined that we would combine these philosophies in a blended approach. Chemists call this a heterogeneous mixture.

Cement trucks spin constantly as they make their way through the city because gravel, sand, Portland cement, and water don't combine into a stable substance until they have cured in their final resting place. Lacking agitation, the water will float its way to the top, and the gravel will sink to the bottom in a puddle of sandy sludge. In *this* heterogeneous mixture of management styles, Chief Sycophant's approach provided the sludge.

I dug right into the work, starting by bringing in a friend and colleague to help me organize and run the whole thing. Clifford took charge of most of the day-to-day building and I concentrated on the dozen or so technically demanding assemblies. A devoted laborer was assigned to help us keep the place in order. She was new to construction but eager and responsible, which beats experience most of the time.

The design indicated a lot of moving parts: a pivoting library wall, foldaway desks, bathrooms with no cut tiles, sandblasted backlit mirrors, and almost every version of swinging and sliding door hardware there is. Then there was everything that wouldn't be seen, the internal systems, which require just as much planning, lest they accidentally end up where they could be seen. Clifford and I spent weeks sorting through all the equipment, hardware, and fixtures, making sure we understood how they would work with the finished surfaces. We spent each day at the desk: He made lists; I made detailed sketches of all the assemblies that were out of the ordinary. Then we'd scurry around snapping chalk lines on the floor to demarcate every surface, cabinet, fixture, appliance, outlet, and door. A pair of carpenters joined us and began with the roughest elements of the job: subfloors, temporary doors, and the like. Our laborer kept the place in order and learned how to open and close the jobsite every day, shutting down the power and water before we left.

Like all New York renovations, amenities were rugged at first. Strings of temporary lights and outlets festooned the ceilings; a plywood-walled toilet was the bathroom; and we washed up every night with a hose that was coiled into a fifty-five-gallon trash can.

A few weeks in, we arrived one morning at the service entrance. The building's superintendent stopped us from going up, telling us the co-op board president wanted to speak with us. We're used to this sort of thing. Construction is noisy, dusty, and unpleasant for the unaccustomed to be around. We get talkings-to all the time, despite the plethora of measures we take to avoid them—serious measures like negative air machines, sealed temporary doors, wetting and covering debris before removing it, sound control mats, limited work hours, nightly moppings, custom removable freight elevator covers, identification badges, and matching T-shirts. None of it works; we still get talkings-to. So we drank our coffee and waited for the president to arrive. He didn't just arrive; he blew through the door of the super's office red-faced and already yelling.

"Fuck," Clifford murmured. He was right.

It had been a destructive night. The valve controlling our garden hose/washing station was original to the building, about ninety years old. Our laborer thought she had closed it off, but it had stopped a few degrees short without her noticing. None of us heard the slow trickle in the bottom of the fifty-five-gallon trash can we kept it in. Hour by hour the trash can filled until it overflowed. It ran for several hours undetected as it seeped through the ceiling of the apartment below. It's not that no one was home; it's just that the water was dripping into a room they never entered: their fur vault. Now, most people don't have a fur vault, so they might not understand just how alarming it is to see water coming from under its door when you wake up in the morning. It is very alarming, fire department alarming, police alarming, lawyer-calling alarming.

In that era, general contractors working on the Upper East Side of Manhattan were required to carry two million dollars in general liability insurance. People used to ask me "Why so much?" The answer is, "Because sometimes you need that much."

Modern Boss was summoned. He met with all the higher-ups and came to chat with us. We weren't fired; we weren't even really yelled at. He had been in this neck of the business for more than a decade; this was not his first flogging. He was a fair-minded fellow. The mistake was a tiny one; the consequences were not. But he could see that no thrashing he would give us could match the thrashing we were already giving ourselves.

Lesson #1: Always double-check everything.

We proceeded with the project; the lawyers and insurance companies worked out the details of the disaster; Modern Boss continued to support us in our progress.

I loved sorting out the technical requirements of the most innovative aspects of the apartment. One area I should have avoided as a neophyte was the fireplace. There aren't a lot of these in New York and this was my first. This was not just an ordinary fireplace. It was small, but the architect wanted the gas to make a single line of dancing flame.

It's fine to be gung ho about taking on challenges, but there are reasons why so many cities' building codes address fireplaces in depth. Fireplaces are dangerous. Professionals make entire careers out of building them. Their hard-won knowledge should be respected. Call them if you want contained fire in your home.

I entertained all sorts of solutions—Bunsen burners, gas jets—and finally settled on a conventional gas-burning artificial log unit that had all the required certifications. I tossed out the fake logs and I customized the burner with a steel baffle that forced all the flames through a long thin slot. The architect came to look at my mock-up; the line of flames was lovely, but could I make them twice as high? Back to the shop I went. I decreased the slot in the baffle to a tiny slit, welded all the edges closed to increase the pressure, and a thin sheet of flames fairly leapt from the thing.

We installed the surrounding steel firebox and finishes. I called Chief Sycophant and asked whether the old wood-burning chimney had been scoped and cleaned during demolition. He assured me that it had. My custom leaping line of fire was complete.

There is so much to do as a job is wrapping up—buffing, polishing, adjusting, and fixing—the last 5 percent is often half the battle. The owners scheduled their move-in; they were an Austrian couple in their forties with two boys, five and seven. They raised their kids following the European model, blithely going about the business of changing households, while leaving the kids in the apartment with our crew as we finished up.

The work was nearly done; we were checking that all the appliances and fixtures worked properly, and it occurred to me that I should check the fireplace as well. Mom and Dad were out of the house, and as carpenters are by nature entertaining, the boys trailed me as I worked, peppering me with questions and getting underfoot. I found the remote-control fire starter, turned on the gas, and gave it a few clicks. Wonderful! The flames leapt in a thin line nearly filling the firebox. They licked the back wall clear to the smoke chamber; it was elegant and hypnotizing. A minute or two in, a distinct roaring sound came from the chimney. I turned off the gas and got down on my knees to have a look.

The smoke chamber was engulfed in flames. What I had taken to be asbestos fireproofing was creosote. The chimney had not been cleaned.

Creosote burns impressively; every internal surface was ablaze. I grabbed a nearby towel I had been using for polishing and climbed into the firebox, my head entirely inside the burning smoke chamber. I smacked the towel against the inner walls, knocking flaming chunks of creosote around my legs and onto the finished walnut floor. Several burns later and with my hair singed shorter, it was finally out.

"Cool!" the seven-year-old exclaimed as I brushed the charred remains back onto the stone hearth. I didn't see any percentage in bribing the kid into silence; Upper East Side kids' silence is far beyond my means.

Lesson #2: Always double-check everything.

After a few more weeks of punch listing, we had finally completed the place. The owners were thoroughly pleased; there were a lot of unusual things to play with and they had all been made to work well. My favorite was a saltwater fish tank that we built into a wall of cabinets in the dining room. Salt water is a finicky thing. It's not just salt and water; there are all sorts of things that go into making it habitable for seagoing fish: pH balance, nutrients, oxygen, filtration, circulation. This tank looked like it was a seamless part of the cabinetry, but there was a whole network of structural supports, filters, pumps, and tubing that we had carefully hidden from view. The cabinet door just above the tank flipped open and lifted itself automatically to simplify feeding and cleaning.

I walked over to admire it one last time before I left. I had never seen it filled with actual fish before. Swimming around happily among several critters I couldn't identify were a ten-inch shark and a lionfish.

As a young man, I watched Jacques Cousteau specials with rapt interest. I wasn't so sure about the miniature bathing suits, but I was certain that scuba diver was one of the best possible job titles. So it dawned on me that it might not be entirely wise to have a lionfish within easy, unlocked reach of five- and seven-year-old boys. That's because it could kill them. They were short, but the dining room chairs were only two feet away.

I went and found their mother. I asked if she knew that the fish was poisonous. She did not; the fish tank was her husband's pet project. She and I spoke about it for some time. I had boys of my own who, left alone for a few minutes, would climb ladders, try out power tools, or throw a cat in the dryer. A lionfish would be far too exotic to ignore. She promised me she would speak to her husband about it. I recommended a cabinet lock and offered to install it myself. I felt slightly better.

As I gathered my things to go for what I hoped was nearly the last time, she touched my shoulder and said, "You know, the boys really like being with you. Maybe you could come one or two days a week and take them to the park or something."

It took me aback. I gave her request a minute to sink in. "I can't really do that. I work every day and I have my own kids to take care of."

"I understand," she said. "I thought it might be nice for them."

For some time after, I thought about that last encounter. I had never seen the father interact with the boys, nor her for that matter, not with much enthusiasm. It saddened me. Still, as role models go, my first choice wouldn't be the guy who destroyed my relationship with my downstairs neighbors and nearly set my house on fire. But then again, she had a point: I would never bring home a poisonous fish for the kids to play with.

Lesson #3: Pay attention to your children. They're not as smart as you think.

BODIES AND BRAINS

Make it snappy.

There is a common misconception that speed and precision are mutually exclusive realms. Experienced carpenters habitually divide themselves into two camps: those who build a lot and those who build accurately. In my early twenties, I heard a story about Michelangelo working at a frightening pace, which, true or not, I have always loved. When I was a child, my parents gave me his biography filled with pictures of his worn and patinaed creations; he has been a hero to me ever since. It pleased me greatly to picture him pounding away furiously with hammer and chisel in a cloud of marble dust while those around him looked on in horror. The image dislodged any notion I may have held that wonderful work needs to move at a ponderous, contemplative pace. I have long believed and have proven to myself over the years that workmanlike expeditiousness is a hall-mark of competence.

People who think speed doesn't matter are fooling them-selves. Industrialists have spent decades in the pursuit of productivity, trying to squeeze ever more work out of ever fewer employees. Robots do the work of people in factories now. Most office workers have jettisoned secretaries in exchange for laptops. Assembly-line theory, interchangeable components, and specialization of labor have permeated every arena of work. General mastery by a single individual is almost unthinkable in most fields. I would guess that there isn't anyone left at General Motors who could build a modern car. Even if one of its executives were locked in a room with every part they needed, a complete collection of tools, and an instruction manual, I would never expect to return one day to find them sitting behind the wheel of a working automobile.

One of the beauties of homebuilding is that, because it began long before technology was even a word, it has remained an isolated bastion of generalism. Until two centuries ago, a carpenter, a mason, or both could provide almost every component of any permanent house found in the world. They had done so for millennia.

The years have chipped away at this model; windows and doors have become more specialized; water was brought inside; gas and electricity replaced candles and lamps; boilers replaced hearths. Things have progressed so far that, in the homes that I build, carpentry and masonry combined might make up only 20 percent of the budget. Architecture, a trade that hardly existed six hundred years ago, now often commands an equal percentage.

The best builders I know still have a broad general understanding of every process that goes into putting up a house. Most of them were either carpenters or masons for a few decades, but as people who love making things, they usually know a little bit about every trade, enough that, left alone for a sufficient time with the right materials, they could single-handedly build themselves a fully functioning house. It might not have programmable lighting and remote-controlled shades, but it would have everything a house actually needs.

I began my career in building by sweeping up and moving things around. Neither of these tasks requires much thought. The only way I could make them entertaining was to find the challenge in them. In the effort to find the most efficient method, I would try all sorts of ways to clean a site. There are plenty of variables in this. What works best? Do you clean from the top down? Start with the center, move everything off the walls, clean the edges, and then reorganize? The problem was surprisingly interesting to me. I liked to see how fast I could do it. In my unscrutinized working world, no one paid me any mind; I was the foreman and the laborer. I would plan my tasks in an orderly way and then sweat myself silly carrying them out as quickly as I could. I developed a few rules for efficiency:

1. Plan until you have a good one. Others are possible, but any good plan will do.
2. Once you have a plan, stop thinking about work. Just work.
3. Do the hardest task, the one you don't want to do, first. Dread is a clear indicator of which task to choose.
4. Do all like tasks together. If you're holding a broom, sweep everything you can, then move to the next task and tool.
5. Without getting anyone hurt, see how fast you can go.

I have carried this simple system through my entire career. To this day, I start by trying to understand the thing I am about to do as fully as I can. I analyze its parts in detailed drawings, plan the steps involved and write them down, measure, calculate, and make lists, ordering every material and component in advance.

Once this is done, I do it all over again, looking for mistakes until I am assured that all my work is as correct as it can be. This is the proper work of my thinking brain. It is a wonderful tool for the job.

But for the next job, the thinking brain is a hindrance. Building things is a largely physical task. The body learns through mimicry, practice, repetition, and refinement. Thinking brains participate in the learning part of this process, which is often slow and frustrating, but once a skill is learned, it belongs to the body; the thinking brain should butt out. Except for the occasional inspired improvement, it doesn't have any business here.

Everyone marvels at the grace of athletes, the way their bodies move as a unified whole, inwardly directed and outwardly expressive. Few of them describe a performance well afterward. That's because their brains didn't do it; their bodies did. I believe everyone is capable of accessing the intelligence of their bodies, learning to appreciate the beauty of relaxed, skilled movements. You can come to enjoy the thrill of doing a task faster and more easily than you thought possible. Sometimes in my career, to see where the limit to speed might be, I've intentionally worked two or three times faster than my accustomed pace. A carpenter friend once told me that he would try to go the whole day without ever allowing a tool's motor to stop running between cuts. The fastest carpenters I've known were the most fun to work with. Friendly competition makes for an interesting workday. If your thinking brain needs something to occupy itself, there's always clever trash talk.

After forty years of working in this way, I have learned a wide range of skills, familiarized myself with all sorts of materials, and where my work meets the work of others, learned to understand the requirements and challenges of their trades. Many of the architects I've worked with, even when reluctant to admit it, know that the experience and understanding that come from physical work are irreplaceable. In the last decade, I've been able to act as a bridge between the cerebral world of design professionals and the skilled and sweaty world of the workers who realize their visions. It can be a creaky bridge at times; there is a lot of animosity on both sides, but it is a challenging, broadening, lucrative, and satisfying place to make a living.

I build for money. That's how I've paid my rent, fed my family, and sent my kids to college. For the last twenty-five years, my bosses have hired me and paid me well because of my productivity. Quality work matters to them, too, but as long as the quality is there, speed is where the money is made.

Not long after I heard the story about Michelangelo, I came across a man who worked with a speed and fluidity that redefined my idea of how fast work looks.

ORLANDO

I love to be near anyone who really knows how to do something. It doesn't matter where they're from or what their work is. It's worthwhile witnessing anyone doing something well. No mastery is easily learned.

When I was a young man living alone on the Lower East Side, I used to walk up to my favorite Polish restaurant, Christine's, to enjoy a big, leisurely breakfast. The waitresses were pretty and impertinent, the place was always bustling, and I was grateful for having nowhere else to be once a week.

Behind the counter was a long sheet-metal grill commanded by one man. Orlando was from somewhere in Central America; he spoke little English, at least not at work, outside of the requirements of his job. He was the first master of his trade I was able to watch work closely. The restaurant had two rooms with at least twenty-five tables, often packed with four patrons each. The whole point of a restaurant is to get as many meals down as many throats in as short a time as possible. Orlando could feed one hundred people a diner menu's variety of perfect breakfasts in under an hour. Waitresses never handed him an order slip, they just gently pronounced some pre-agreed code and listened for his "Coming!" before they bustled off confidently.

His movements were relaxed, minimal, fluid, and nearly constant; he never appeared to think or calculate; he just cooked and cooked and cooked. While waiting for the grill to do its work, he replenished supplies, cleaned bare surfaces, fetched more dishes, and prepared the simple garnishes he added to each plate.

I am always impressed by people who can do things that I can't. Sure, I can rustle up an omelet, but here was a man who was making a symphony of short-order cooking. Sitting at that counter, I learned more from Orlando about the expressive potential of work and the beauty of its mastered movements than I have from a string of paid instructors.

For a while, I thought I wanted to be a cabinetmaker. I'd read books and magazines about furniture building that were filled with romantic musings on craft, the natural beauty of wood, and the noble life of the gentleman woodwright. I composed a mental picture of a rose-rimmed house in the country, a sturdy outbuilding beyond it, the windows aglow with the internal light of industry.

In pursuit of this vision, I found a newspaper ad that led me to a shop in Brooklyn's Clinton Hill. I climbed the subway stairs to a leafy brownstone-lined street. The sidewalk's flagstones rang with dreamy promise. I found the street the shop was on and followed it as the numbers ran lower. With them went my expectations; the flagstones gave way to concrete; the brownstones petered out and the expressway appeared. The street ran downhill and under the elevated roadway to a curdled-cream-colored industrial building on the artery's darker side.

A few flights up a concrete stair, an antique metal fire door barred the entrance to the workshop. I pushed an old doorbell and a scraggly-haired, half-bald man several inches shorter than me appeared and barked like a terrier with a Louisiana accent, "Yeah?"

"I'm here about the job ad in the paper."

"Follow me."

He led me to an enclosed office where a thirtyish lanky Irishman with flushed cheeks wrote down my bona fides. He showed me around the shop, stopping at each machine to ask, "Do you know how to work this?" At all but the table saw and the drill press, I lied, "Yep." I was hired on the spot.

For the mid-eighties, the shop was a sophisticated operation. It had European sliding panel saws, edgebanders, a line-boring machine, a thirty-two-millimeter carcass joining system, big shapers with automatic feeders—all the accoutrements of a modern millwork shop. Most of it was completely unfamiliar to me, but it didn't take me long to figure out that this place was geared up to make any size box with a door you might want and that was about it. There was no hint of traditional joinery here, no careful honing of chisels to make perfect mortises, no smell of spirits and French polish in the air; this was a factory. I was assigned a workbench next to the most senior cabinetmaker in the shop and put to work in the "mill."

The mill was the area of the shop where rough lumber is turned into square, dressed, dimensioned stock of whatever size is needed. I had a laborer show me how to work the jointer and thickness planer and did a passable day's work. As soon as I got home, I pulled out my books and magazines so I could learn how to do what I had just done that day. I hadn't gotten it too wrong. Each day I would do more and different things, and each night I would study up on the theory behind it, until after a while I pretty well knew my way around.

There were a few rough patches. The sliding table saw was nothing like the simple stationary tools I grew up with. It was wonderful at cutting panels into perfect rectangles, but it was awkward to stand at. Cutting smaller pieces, I never knew how to place my body without overreaching or worse. One day I tried to trim a small panel by standing to the right of the sliding table, pushing it directly from behind. The piece bound between the fence and the blade; the saw's nine-horsepower engine threw the piece into my chest with such force that it knocked the wind out of me. Thanks to my brother's post-scuffle instruction years before, I knew how to restart my breathing, but it was a struggle to pass the event off nonchalantly.

Worse still was the edgebander, a whirring collection of feeders, glue pots, applicators, pressure rollers, trimmers, and loppers. There was so much to the machine that could go wrong that it was cause for celebration on the rare days that passed without edging disaster. I'm still not sure if I could make that thing work.

After a year, pretending had transformed into monotonous proficiency. There was less and less to find interesting about the place. After a thousand or so, a box is a box, and a door is a door; variations in size and wood species do little to reignite the fires of fascination. I would content myself with finding improvements in process. My cut lists became more complete and finally comprehensive; the clever placement of carts and workpieces organized and sped the flow of my day; joining and assembling became rapid and thoughtless; sanding was systematic, complete, and routine. I refined myself all the way to utter boredom.

One Monday, my red-cheeked manager came to me with a "production job." An office build-out required ninety-six cabinets in three sizes, components of identical workstations. He showed me the drawings and told me he needed it in two weeks. For my helper, he assigned a newly hired, fresh-faced, freckly young woman with curly black hair who had an attractive, loose-fitting way about her. It was probably a crush, but I passed off my eagerness to show her the ropes as benevolent seniority. I walked her through my system for writing cut lists, some of it borrowed, some of it my own, that showed every operation in the order in which it was performed on the saw, maximizing the yield from each panel, and indicating on which cart each piece would be placed when finished. She took to it like water.

We set to work like a pair possessed. Each morning we'd meet, set our goal for the day, and burn through the stack of panels and parts until every preparatory cut, hole, and mortise was done.

Assembling was the final step, and the most complex. We started on Thursday morning, learning the little dance that each cabinet required, gaining assurance as each completed box left the bench. By the end of the day, we had forty cabinets assembled with doors attached. We were assembling one every twelve minutes. That was fast. So the next day we started timing each cabinet, always trying to beat our last record. By the time we got to the last one, we had it down to seven minutes. It was beautiful, compact, relaxed, wasteless, and liberating.

We finished the remaining fifty-six cabinets one hour before quitting time, and the entire project in under half the time the shop manager had allotted for it, and we knew it. I escorted my helper into the office's conference room and made a pot of coffee. We sat for the rest of the day in full view of the manager, shop foreman, and other cabinetmakers. We slurped our coffee, chatted about life, talked about goals, and enjoyed our fresh accomplishment. No one said a word to us.

Our conversation reverberated in me as I made my way home that night. Most custom cabinet shops do a few things well; they all make boxes and doors, adding a few bells and whistles that justify their expense. Kitchens, bathroom vanities, and shelving units are their bread and butter; modern shops aren't tooled up for much else. I had found some room to learn in the variations of door styles and finishes, but not a lot.

I had just spent the week turning my helper and me into a miniature assembly line, and I had to admit to myself that in most ways we weren't craftspeople—we were factory workers. There is a remote corner of woodworking in which a select few artisans make one-of-a-kind, gallery-ready creations. Most of them struggle to pay the rent. With each step of the way home, my vision of a romantic future in woodcraft evaporated.

That weekend, I searched the papers for a new job. Cabinetmaking wasn't my thing anymore.

That's the conundrum of competence: All that effort, everything that goes wrong, countless corrections and solutions finally come together, and for a glorious while work is a banquet. But the next day comes, and sooner or later a body needs breakfast.

+ + + +

CHAPTER 8

Tolerance

Ugh! People!

About a year ago, I got a call from a man who asked if he could interview me for a magazine. I misunderstood the name of his publication and was indifferent to the idea, but he was affable and sounded genuinely interested in the New York fancy building world, so I agreed to meet him near my jobsite for a drink after work. During our conversation, I figured out that he worked for a magazine of better repute than I had supposed. I'd never been interviewed for anything. He asked incisive questions, seemed to have done his research, and was genuinely fun to talk with, so I agreed. Fifty hours and five months of interviews later, he determined that we were done. The article came out a few months after that. It was long, but pleasant enough to get through. He gave me most of the punch lines and managed to steer clear of the most derisive things I might have said. I suppose, then, that it was a success.

One topic he revisited several times over our months together was the idea of tolerance. Perhaps someone else had spoken to him about it. I would show him the layout lines on the floor of my job and say, "The cabinets will go right there." He kept asking, "How much can they be out? What is your margin of error?"

"They go right there," I told him. "That's what the line is for."

Tolerance can mean a lot of things. One type of tolerance is the quality some people develop that keeps them from acting maliciously toward others when they feel others have behaved imperfectly. It is a polite halfway station between hatred and compassion, designed for the purgatory bound. The tolerance my writer was asking about can be described as the level of imperfection in an item that will be put up with before a customer is no longer willing to pay for it.

This is a hard needle to thread. One could consult the customer: "Mr. Bigwallet, how warped can we make your doors and still count on prompt payment?" That approach probably wouldn't work. So fabricators form councils and guilds to discuss the issue.

MEMBER: Every time I make a bunch of doors for a job, no
matter which method I use, at least one of them warps
about an eighth of an inch over eight feet. I'm sick of architects rejecting them. How many centuries will it take
for architects to realize that wood is an organic material
that moves?
GROUP: Danged if they ever will.

As a remedy, they agree to write a book of reasonable standards stating that custom-made doors shall have an allowable warp of up to one-eighth of an inch in eight feet. In all future contracts, the members cite the *International Woodworkers Guild Book of Standards* as the arbiter of quality for their industry. They sleep comfortably in the knowledge that, from now on, when that one door out of twenty gets too much moisture on the bathroom side and warps, they will still assuredly be paid. They react with dismay when, six months later, they get a call from a contractor: "We got a problem with one of your doors over at the Bigwallets' job. When can you take a look at it?"

One by one, each of the Woodworkers Guild members learns that the Mr. Bigwallets of the world don't care a jot for their *Book of Standards*. His bathroom door is warped, and he wants it fixed. As correct as craftsmen are about the unforgiving nature of wood, they are mistaken about the magnanimity of the wealthiest among us. Few are willing to defend their rectitude in the face of an unsatisfied customer who makes one thousand times their salary.

This state of affairs remains just as thorny as it has ever been. Industry standards don't provide a useful guide to building things people will love; they're just talking points for lawyers to follow when they fight about the things people hate. When work gets rejected, citing industry standards might help with arbitration, but at that point things have already gone too far.

It is possible to make one's way to results that satisfy all parties. I ask myself three simple questions when I build:

1. Does it look right?

   "Can't see it from my house" and its corollary "Good enough for someone else's house" are minims that a surprising number of workers spout with regularity. In practice, we never get away with anything. If I can see that something isn't flush, or centered, or spaced regularly, the client can see it, too. If I can feel a misaligned joint in a banister or countertop, then it's not right. That means it will have to be fixed.

Some clients are difficult to care about; that is not our concern. There are other reasons to build things as accurately and thoughtfully as one can: It's easier to build something right than it is to build it wrong and then fix it. It's satisfying to build beautifully made things, especially things that the maker finds beautiful. Being paid readily, without complaint, is serene.

Steel workers hoist half-ton beams and bend thick plates into curves on enormous hydraulic presses. They can be forgiven if their work doesn't split the thickness of a pencil line. Most of what they do isn't seen. It's easiest to assume that they will build with some slop to their work and to plan, in advance, how the gap between their imperfect work and the precisely laid finished stone slabs it supports will be bridged. This same idea can be applied to every step of construction. I always allow for a margin of error and a way of mitigating it so that the finished surface can be placed exactly where I want it, right down the middle of that original snapped line.

I have heard contractors repeat pablum like "Make it beautiful, even where only the angels can see it." In my opinion, they are half right. Work should be neat and organized every step of the way. This inspires confidence in the client and morale in the building team. But taken too far, this noble-sounding attitude leads to the obliteration of the construction schedule. Clients care deeply about schedules. The only part of the finished product they really care about is the part they and their peers see. That's the part that makes the neighbors look around and say, "*Damn.*"

2. Does it work easily and reliably?

People love things that feel right. Better still if they look great doing it. Engineering and machining provide the feel. Shape and finish provide the look. If a mechanism grinds, binds, wobbles, takes too much effort, is too complex to operate, or needs frequent repair, clients will not like living with it. It doesn't matter how sexy it is if it doesn't work. I've heard wealthy folks curse the Ferrari when it takes one too many trips to the repair shop. Very few customers know or care about the satisfaction of press fitting a bearing into place. But anyone can hear the squeak it makes if it's not done right.

3. Is anything about it annoying?

In some ways, this is the easiest standard to apply. With great reliability, most people prefer: smoothly arcing curves, precisely aligned components, connections that look like they will hold, switches and controls arrayed attractively in locations they can find without thinking— lights that don't shine in their eyes, speakers they can hear clearly, ugly things concealed, drains that don't dry out and smell, doors that open effortlessly, stationary things that don't come loose, touchable things that aren't sharp, rough, jagged, or slimy. We are raised with varying but ubiquitous capacities for annoyance, and we are an unforgiving species. If a rule of thumb is needed: Ask an eleven-year-old for their aesthetic opinion. If they like what they see, you'll probably get paid.

BLUE TAPE

Seven years ago, before I gave up working for contractors for good, I took a job as a project supervisor for an up-and-coming outfit from Long Island that was establishing a foothold in the city's high-end market. A friend introduced me to them, and I got the job without much effort. Little was asked about my history in the business, and from my side, little was volunteered. My two-decade marriage was coming to a close; all I wanted from work was to meet the daily needs of my boss and to go home at four-thirty each day. An entire career can be navigated this way. I was only looking for a single year's respite.

The man who hired me was fifteen years younger than me, six inches shorter, and 120 pounds lighter, so I took to calling him Little Boss behind his back, not as an insult, but because anyone seeing us together couldn't help but notice the contrast.

After a few months' tryout, Little Boss gave me a project of my own to run, a Park Avenue duplex that was about to be gutted and completely redone.

The apartment had been renovated twenty years before by the very outfit that built the Snail job. Had I stayed on at that company, I would probably have been part of the crew that built the apartment I was about to demolish. It was an elaborate place, heavy with classical moldings, carefully shaped marble, and trompe l'oeil painting that had been out of fashion for more than a decade. The work had aged well, even if the style hadn't. Doors still hung neatly in their jambs, joints were tightly fitted, and the paint job was still crisp.

As we ripped it to pieces, I recognized graffiti drawn by men I had worked beside years before and imagined my own work suffering the same fate.

There was little time for rumination; this was a summer rules job requiring all the focus I could muster. For three scorching months, we toiled to bring the apartment within shooting distance of completion in its second year. By Labor Day, none of us knew whether we'd be back for one more summer or two.

Impending divorce was commanding much of my psychological wherewithal. It made work dreamlike, not that it was less engaging or demanding, rather less substantial. From years of habit, I could bring my attention to the job, solve things that needed solving, and direct my crew, but it all felt like play-acting when contrasted with the drama of my inner life at the time.

After the summer at the Park Avenue duplex, I was assigned to an eight-month project in Greenwich Village, a cultured heiress's do-over of a townhouse that mostly needed a new master bath and a better paint job. Attached to the project was an owner's rep, of whom Little Boss was noticeably afraid. She was a whirlwind of pointed questions and jargon-filled declarations who didn't seem to think much of us builders. But work was my refuge now and I was very much on top of things, even if I wore it loosely. She seemed to appreciate my offhandedness, having accustomed herself to unction, so she quickly softened and became downright friendly. She enjoyed being respected without being feared.

The townhouse had been recently renovated by a construction firm no one had heard of, under the direction of the same architects we were working with on the Park Avenue duplex. It became clear after days working there and probing its rooms that the previous renovation had gone very wrong. Our first evidence was the air-conditioning system. I found ducts that led to nowhere, grilles that were incapable of drawing in or expelling enough air to cool the oversized rooms, and units installed so that they were unserviceable.

We tore out drywall and fixed failures as we found them. I started to keep my eyes and ears open for trouble.

Next, it was the roof. Whoever had installed it had done so in reverse order. Instead of water being directed from upper layers to lower ones and downhill to the drainpipes, it was directed from upper layers *under* lower layers and on through the joists to the ceiling below. The entire roof was covered in an elaborate but poorly supported deck. We tore it off and replaced the one-year-old roof in its entirety.

Soon winter came and a pipe burst in the nearly new kitchen. No one had bothered to insulate the kitchen's walls . . . out came the cabinets.

A few weeks after that, someone dropped a bag of concrete on the floor where I had set up my office. The whole room shook so profoundly that I knew something was terribly wrong. I opened the ceiling around the staircase where the main structural joists could be found; they were cracked nearly through. I traced them to the central bearing wall between the stair hall and the parlor. The entire five-story wall that ran the length and height of the townhouse, supporting every floor, had been demolished and replaced with light-gauge metal studs that had only enough strength to support their own weight.

All of this work had been done by the contractor who had performed the previous renovation under the watchful eye of one of New York's top architecture firms. The property had no outstanding violations, so everything had been inspected by the Department of Buildings and passed. This townhouse was in the heart of Greenwich Village, one of the city's most expensive neighborhoods. Its new owner, my client, had plowed tens of millions of dollars into a house that could have collapsed completely under the weight of a slightly too extensive sculpture collection.

It fell on me to tell the owner's rep that a New Year's Eve dance party was out of the question. The entire central wall needed to be replaced from the basement to the roof, re-supporting all the floor joists along the way. My owner's rep still appreciated me, but she seemed to wish I was less conscientious.

After that, I became reluctant to take a hard look at anything; all of our predecessor's work was bad and all expensive to fix. The sump pump in the basement was wired to electrocute. The chimney in the roof's private study caught fire when I lit the gas logs; it had been built with flammable materials. We ordered smoke tests on the fireplace in the master bedroom. It exhausted through the flue of the bedroom just above it. A romantic evening around the fire for Mom and Dad would have meant waking up the next morning to find the kids asphyxiated in their beds. The simple little paint job and bathroom ballooned into a fifteen-month owner's nightmare. Summer was coming and I had to get back to Park Avenue. I handed the job off to an old friend and wished him luck. He didn't thank me.

I had been preparing for Memorial Day on Park Avenue as early as January. My project manager and I made a list of everything that needed to be completed and worked backward to where the project currently stood. The most alarming things were the ones that weren't even designed yet: a steel and stone staircase, the kitchen, and an elaborate master bath/dressing room/sitting room suite.

Contractors know that our business is all about kitchens and baths; they require the most trades, coordination, and schedule sequencing. I know, from years of building them, that staircases are the next hardest thing to complete. We drafted a registered letter to the architects and owners declaring that these aspects had to be designed by March if we were to stand any chance of finishing in a second summer. That gave the architects another six weeks to get us designs. They had already been designing the job for three years.

A sensible person might think that a project we had been working on since the previous summer's start, especially one with such draconian deadlines, would already be thoroughly conceived by the time we began. They would be mistaken. Three decades ago, I regularly saw thick sheaves of hand-drawn blueprints clearly showing the architects' intents, but for the last twenty years, all I have seen are thin booklets of computer-generated stick drawings with nary a buildable detail to be found. These are usually accompanied by an explanation that this is the architect's "fast-track" method of building and that detailed drawings will arrive imminently. I take this kind of talk as a sure sign that the project in question is going to be a long haul.

The registered letter was written mostly for its value as a paper trail so that, come summer's end, with the inevitably incomplete kitchen, master suite, and stairs necessitating another year's delay, fingers could be pointed in the right direction. It had no noticeable impact on the architects. By May 1, we didn't have usable drawings for any of them. Unsurprised, we sat down to write more registered letters stating that things weren't looking good.

Summer came. I called in all the reinforcements I could find. I was aiming for a replay of the East Seventy-second Street job; I meant to get this thing done.

This owner irritated me. He would feign fellow feeling, sidling up to me to ask conspiratorially if his place would be finished on time, just between us. I pointed out that the letters we had sent clearly stated that it could not; the architects had not delivered what we needed when we needed it, so the schedule was ruined before it got started. Bad news cannot be repeated enough. I have delivered it by every modern means of communication available, sometimes all of them combined, and it almost never finds its mark. Repeated experience has taught me well that no amount of meeting-minute distributing, registered letter writing, urgent emailing, or face-to-face explanation will deflect the blame heaped on a contractor when a Park Avenue owner doesn't move in on time.

It wasn't only the architects who conspired against completion: The carpentry team Little Boss had hired was not up to the task. Few of them knew anything more than the basics of framing and sheetrocking. Even worse was the stone and tile company. What they lacked in skill, they made up for in mismanagement and lying. Their own tile setter quit twice in fits of pique. Their project manager was argumentative at first, followed by unreachable. Only later did we learn that he was serving time in Colorado for a real estate scam.

On the plus side, Little Boss let me have all the help I needed. I hired one of my closest friends, Pauline, to be my project manager, and she brought an eager assistant super, Jesus, a lucky name for a project in need of a miracle.

We musketeered the place for three months. Pauline ran the purchasing, Jesus ran the crew, and I made sure the subcontractors built everything fast and right. The owner kept up his weekly sidles, always eliciting the same answer, right up till the Friday before Labor Day.

Pauline has a way with people. When the co-op president and resident manager came to inspect, we weren't really done, but they ruled it close enough, mostly because they liked her. Little Boss saw that we musketeers were out of fight, so he sent us away to run a new project, offering it as an opportunity to recover.

A fresh team was brought in to finish what little remained to be done. Little Boss said nothing about what we had accomplished, neither appreciative nor otherwise. The owner didn't bother thanking us, even though we saved him a year's Park Avenue mortgage and maintenance fees.

Our success in the Park Avenue frying pan must have convinced Little Boss that the three of us were ready for a full-fledged fire. Our new project was in a nearly completed Upper East Side high-rise with leaky windows and a substandard air-conditioning system. It had been hastily fitted out with kitchens and baths to fulfill the Department of Buildings' habitation requirements. Several tenants had already moved into the lower twenty-six floors. The top six stories, totaling twenty thousand square feet, were ours to finish. They were being combined into a single home for a nameless hedge fund owner. The top four floors were for his family, and the lower two were meant to be a sort of guesthouse for his in-laws.

By the time I got there, the unused kitchens and baths had gone to the landfill. Developers only install them to get a certificate of occupancy from the city. They are built with the types of appliances middle-class homeowners use in their homes. I don't know a single architect who would consider reusing them. There's no commission in it. We throw them out because the resale value of "used" mid-grade appliances and fixtures can't pay for the time it takes to find them loving homes.

When I arrived, the only real work going on was an endless effort to seal the windows as rainstorms, blowing in from each new direction, revealed their breaches. Activity picked up a few weeks later when the rooftop air-conditioning chiller spilled its entire antifreeze-filled contents into the inhabited apartments of the bottom twelve floors. It was the first time I have seen apartments covered in actual inches of antifreeze. The damage was riveting and, thankfully, not ours to repair.

On *our* project, an English design firm, coasting on the name of its dead founder, was in charge of aesthetics. In order to meet local codes, a New York architecture firm had been engaged, largely for the value of their official stamp. The head designer was all of twenty-seven. He was admittedly a lovable young fellow, a bit giddy from his position, but humble enough to know when he was out of his depth. His assistant was younger still. She was kind, funny, and enjoyed the conviviality of the jobsite. They were overseen by a senior adviser in the British nanny mold, who alternated between flirtiness and indignation. She liked to show that she was in charge, but her authority evaporated early on when she took an ill-advised trip to Greece and purchased several hundred thousand dollars' worth of quarried marble block with the clients' money but not their permission.

My team consisted of Pauline, ever my right hand, whom I met as an office assistant and who was now a seasoned project manager. Technically, she was my supervisor, but she never would have said so. One of my oldest friends, Clifford, joined me again as job super. He and I had met in the mid-eighties as carpentry pups. We pushed each other for decades to be faster and more skilled than either of us would have been alone. Jesus was assistant super. He was the best at interacting with the crew; he enjoyed authority but lacked the impulse to abuse it. Little Boss assembled us, announced the importance of the job to the company, and told us we had eighteen months to finish it.

Little Boss had already chosen most of the subcontractors we were to use. Two of his choices were instantly alarming. Inept Marble, from the Park Avenue job, was selected to do all the stone and tile work: three and a half kitchens and eighteen bathrooms. Their project manager had been recently released from his Colorado prison cell. He was eager to get back to work.

This prospect was nightmarish, and I said it so many times

that a meeting was called with Little Boss, my team, and Inept Marble's top management. The meeting was as useful as most; patronizingly intoned reassurances were offered that the difficulties of Park Avenue had been addressed, taken to heart, and thoroughly remediated. Indeed, we could expect to see a whole new Inept Marble right from the get-go. Little Boss seemed confidently convinced. My respect for him fell off a cliff.

The millworker he chose was no better, though it wouldn't have mattered who was selected; there is not a single millworker I know of in the city who could draw, produce, and install nearly ten million dollars' worth of cabinetry, doors, and trim in the allotted time. By contrast, the East Seventy-second Street Summer Rules job, with one-eighth the budget, was done with five millworkers. There are shops that can produce enormous volumes of work, but none that can draw or install it that fast. With utter confidence in their ability, and a well-founded belief in the quality of their woodwork, Little Boss chose the shop that his father had founded more than forty years ago, and still ran.

How does one tell the boss that his father, hardworking and skilled as he might be, is not up to the task? In this case I said it over and over. Every step of the way it was clear that we were losing ground. The woodshop assigned only one draftsman to the project. With each sequentially drawn room, the schedule drifted away. The main kitchen, which had more than a million dollars' worth of cabinets, wasn't even drawn until the originally scheduled due date for the entire project had passed.

But that wasn't the real problem. The real problem was that we were trapped between loyalties and worlds.

Little Boss's father had immigrated to the United States from western Europe decades earlier. He started a woodworking business and later a general contracting business that focused on quality, good customer relations, and honesty. The company relied on a workforce made up largely of his compatriots. Decades later, when I worked with them, many still spoke little English. They were an insular, old-school European community who lived by harsh rules.

The installers were openly afraid of the father; all of them owed him their livelihoods, substandard as they might be. More than once, one of them would tell me, "The Old Man doesn't want it done that way." When I would answer, "I don't care how he wants it done; he's the cabinetmaker; he's not running this job," his workers would look at me in horror and disbelief. They held large family-style Christmas dinners with home-cooked favorites from the old country. The arrangement was charming and draconian by turns.

Like the workers, Little Boss was fiercely loyal and equally frightened where his family was concerned. He shared his father's focus on quality, but unlike his father, he lacked the hands-on experience to understand what it took to achieve it. He had spent some time working in the shop, but more time going to college studying the business and management side of building. He believed in meetings, walk-throughs, documentation, and the promise of the digital world.

As the weeks slipped away, the friction between us grew. Inept Marble performed exactly as expected. Room after room of this project was filled with marble walls, sinks, benches, and ceilings, all of it interacting directly with cabinets, lights, glass doors, and electrical devices. No subcontractor can be expected to know where everyone else's work is supposed to go, and architectural drawings never show it correctly, so I always assume it is my responsibility to make sure that all the dimensions on a subcontractor's shop drawings are correct. I spend hours checking reference lines, plumbing layouts, millwork locations, and whatever else is in a room. I carefully annotate the subcontractors' shop drawings in red ink and send them back to the drafter for correction. Then I check the corrected drawings all over again. Once I'm positive they are right, I release them for fabrication. In a job with three and a half kitchens and eighteen bathrooms, it is a herculean task. I found it impossible to devote every moment of my days to monitoring companies that couldn't keep up.

The upper levels of the house had a four-story fiberglass composite staircase, fully paneled dining room, living room and entryway, TVs hidden behind motorized panels, a heated roof deck with sliding fire feature, another roof deck with a custom-made hot tub complete with a TV that automatically arose from the deck, and marble outdoor gas fireplace; the list of over-the-top fanciness is so long it gets tiresome. Keeping all of it going at once was beyond trying for our little team.

I managed to establish a rhythm with Inept Marble. One of their installers could be reasoned with. I spent untold hours with him, showing him how to lay out, correcting each of his drawings; it was painfully slow, but the stone slabs arrived properly cut, until it came to installing the master bathroom. For reasons I will never understand, every slab in that enormous room was cut to the wrong size. Nothing was even close. Twenty-five slabs of hard-to-find stone were wasted.

Everyone was upset. Little Boss called a meeting. Inept Marble's owner was there, the Colorado scammer, a handful of other managers and installers. Pauline and I stuck close to each other. I brought along the shop drawings they had originally made and on which I had corrected nearly every dimension neatly in red. I brought the subsequently corrected shop drawings, bearing my signature, which showed the dimensions matching my corrections and approved for fabrication. I led the party to each location and showed how, if the corrected dimensions had been used, every slab would have met millwork, plumbing, and fix-tures in just the right spot. Then, I showed them that each slab had been cut to a size having nothing to do with those dimen-sions. The Inept team harrumphed and made conventional ex-cuses. After a long discussion, some of it heated, they agreed to remake everything at their own expense.

It wasn't much of a victory. Afterward, Little Boss took Pau-line and me aside and told us problems like this were the result of "lack of management." I can be argumentative at times, but I had nothing to say to that. I had done everything short of going to the stone shop and cutting the slabs myself. My opinion of Little Boss, already at the bottom of a cliff, began a slow wade into the freezing lake at its base.

From there, things deteriorated further still. We were months past our due date. Every week, Little Boss would come for his walk-through and ask, "Why isn't this room done?" I would answer, "Because the millwork shop hasn't delivered the doors yet." Or, "I haven't even got shop drawings for these cabinets." He would assure me that he would speak to his father, but noth-ing changed. Little Boss brought in his senior super to help speed things along.

Senior Super had worked for this company for twenty-six years; he spoke the language of the millwork installers, and I respect him greatly, but even he was unable to coax any speed out of the father's crew. One day, after the walk-through, he pulled me aside and said, "Mark, you can't keep saying that things aren't done because of the millworkers."

"But he keeps asking, and that's the reason."

"I know, but it's his father's company. Now it's his company. He feels like he owes everything to the Old Man. When you say they're doing a bad job, he can't hear it; it upsets him and it makes him mad at you." He was a smart man, but I still didn't know what to do. The next time we walked through and Little Boss's "Why isn't this room done?" questions started, I would answer, "Uh, there's no door." It seemed ridiculous and surreal, but it kept the peace.

A year after the due date, we were finally wrapping up. The designers were preparing to move in furniture, while we ran around punch listing. The agreement with the designers was this: They would go through every room in the house, mark anything that needed fixing with a piece of blue tape, and only when we were done with the entire punch list would they allow any furniture to be brought in. They rejected the idea of finishing floor by floor, decorating the apartment as we did so. Who could think that was reasonable?

This was a convenient way for them to blame us for any and all delays. We found out through the friends we had made among them that much of the custom furniture wasn't finished and they needed to buy time. We'd give them a room to inspect, they would speckle it with blue tape like Smurf smallpox, and we would bring in a touch-up crew to "fix" it.

I know several really good touch-up artists. They can make almost anything disappear—cracks, scratches, smudges, chips—in just about any material or finish. The difficulty here wasn't fixing things; it was figuring out what we were supposed to fix. All day long, the artist would call me in, show me a cluster of blue tape pieces, and ask, "What am I supposed to do here?" I'd take out my glasses and sometimes grab a work light to inspect the spot. Fully half the time I couldn't find even the slightest scratch.

The next time Little Boss came around I showed him the situation. "Our job is to make it perfect," was his only response. I gave up on any help from him. I told the artist to do everything she could see, and when she couldn't see anything, to peel off the blue tape and throw it out. When she was done, we invited the designers in for one last inspection. To rooms that once had hundreds of little blue dots, they might add five more. They couldn't see anything either; they were just making work and running the clock to avoid accountability.

After a few more weeks, the furniture was in. The clients would be arriving soon. They seemed generally satisfied with the process despite the egregious delays. Little Boss didn't come around much; he was focused on another major project where he was trying out a new management methodology, one in which he placed the highest of hopes.

One afternoon, his chief project manager came by. We had become friendly over the course of the job. He wanted to tell me about the new direction in which Little Boss was trying to take the company. He had invested in an outfit that did 3D modeling of entire jobsites, all the systems behind the walls, hardware, everything. The project manager wanted to know what I thought of the idea. It turns out he didn't want to know once I started answering, but I told him anyway. We were sitting in the simplest child's bedroom in the house. I walked him around the room and showed him twelve different assemblies that were drawn incorrectly in the two-dimensional computer-generated drawings the architects had provided:

1. The pockets at each window where roller shades were to be concealed had strip lights with no shields. These lights were meant to be hidden, but at night you would be able to see them in the reflection of the floor-to-ceiling glass windows. They would have looked like bargain-bin Christmas lights. I had completely redesigned the shade pockets throughout the job to be attractive in the night-time reflection.

2. The floor outlets the designers had specified wouldn't fit

between the wooden finish floors and the concrete building slab beneath them. I had custom outlet boxes made
using UL listed parts from other assemblies so that they
could fit and still be legal.

3.–12. Et cetera.

I went on for some time, until he grew aggravated: "We had
2D CAD drawings on this project that were almost useless when
it came to anything specific. Do you think that a bunch of computer whizzes, with no building chops, sitting in an office somewhere would have caught any of this? All you're going to get is
3D models that are just as screwed up as the 2D models, but
with another dimension added. Someone who knows how to
build has to be here, think things through thoroughly, and solve
these mistakes one by one. No one working in some office
somewhere even knows there are, quite literally, hundreds of
problems here. I just showed you a dozen, and this was the easiest room in the house. Want to go talk about the staircase?" He
demurred. Too bad, I could have given him examples for days.
I had drawn hundreds of sketches that illustrated each and every
one.

A week later, I called Little Boss and told him I wasn't needed
on the project anymore; the punch list had dwindled to the
point where they could finish without me. "Where am I going
next?" I asked.

"Oh, uh, we haven't really lined anything up for you."

"So you're telling me I'm out of work tomorrow?"

"Yes."

"You weren't planning on giving me any notice?"

"Someone was going to call you."

"They didn't. I called you."

"Oh, sorry, we really don't have anything."

I hung up the phone, enraged. I stayed there all morning ignoring everything around me. I was busy calling the people I trust most in the business. By afternoon I had a new job, so I left.

That was the last time I worked for a general contractor.

I never saw or spoke to Little Boss again. He had been putting all of his time and attention into a project that had been running parallel to mine. He had intended it to be a proving ground for his 3D-modeling venture, an industry-changing innovation in which he was an avid early investor. That project was roughly equal in price to the one my team worked on. It started about a year before ours and finished two years after it. The thought still brings me glee.

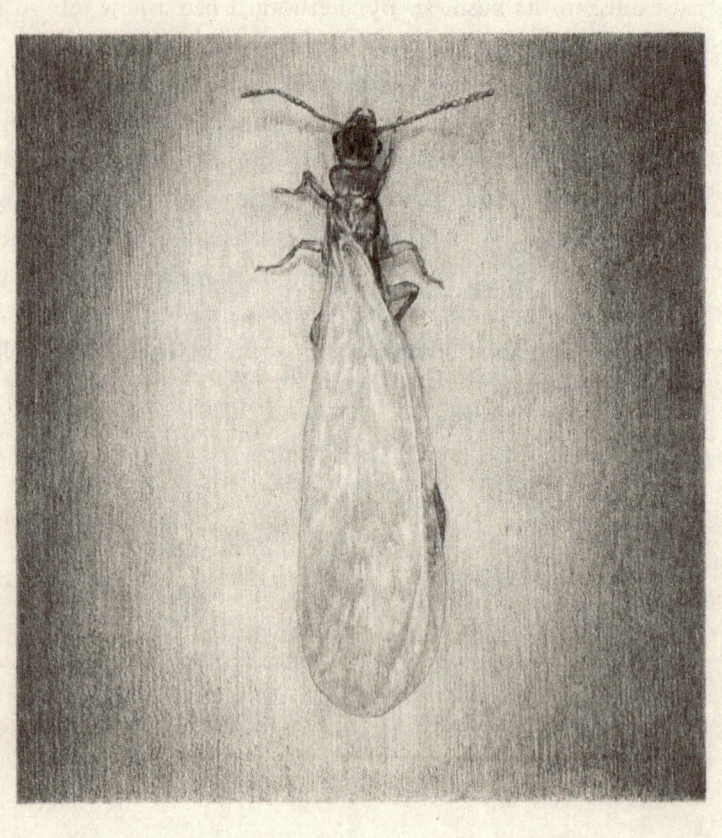

CHAPTER 9

Fear/Failure

"It can't be done; it won't work."
—*ANDY THE CABINETMAKER*

Fear is a fascinating mechanism. It's designed to protect us. When danger approaches, it releases powerful chemicals in the brain. In some instances, these chemicals can cause complete paralysis, as in the case of the possum that mimics a corpse; in others, they become performance-enhancing drugs that enable a mother to fight off an attacker who would otherwise overpower her. There's nothing wrong with the mechanism itself, except if it inhibits our ability to act when there's something we wish to do.

Even in mildly frightening circumstances fear slows the brain and sets it to calculating everything that could go wrong, complete with visions of embarrassing or painful outcomes. The next thing you know, you've scratched on the eight ball or tumbled off a cliff.

Many try to pass off this slowing down as deliberate, calling it thoughtfulness or care. I've seen carpenters spend hours pondering an iffy process long past the point when they should have just given it a try. Usually all that pondering only increases the pressure and worry; they'll find out soon enough whether they have the skills or not.

Any fighter knows that one of the most difficult aspects of fighting is learning to relax when things get the most heated. They train the same movements thousands of times so that when immediate reactions are required training takes over where trepidation and thinking fail. After countless iterations of a technique, faced with an opponent, an action either works or it doesn't. There's only one way to find out.

In my line of work and many others, people who are in a position to do so routinely use fear as a weapon. Most interactions I see between men are based on intimidation, especially the initial ones. It may be cultural, it may be animal, but men generally go through a ranking ritual when they meet: Who is in charge? Who is the scarier one? Who could do the most damage to the other? Once that has been established, the relationship progresses, usually as the hierarchy would predict. Business relationships, friendships, and loves are all colored, at least in part, by fear. Every client I've had could destroy me financially for sport—what a precarious relationship. It has taken me many years to identify violence, in others and in myself, as the root.

I was raised by self-professed pacifists with little experience of physical aggression. As a young man, I took up boxing, then tai chi, and finally traditional Chinese kung fu explicitly as a means of understanding violence and intimidation in my relationships. I worked at it for thirty years. It has helped to be hit, and hit hard, so that now I know not only will it probably not kill me, but I don't even mind it much. In a meeting with someone accustomed to bluster and bullying, just knowing that has a transformative effect on the dynamic. I don't think this is something that can be taught by a book, unless someone hits you with it repeatedly.

More internally, fear keeps us from admitting what we are. There is so much in us that doesn't like to be looked at; we think that if our friends and neighbors found out about it, they would drag us to the town square and throw us in the stocks. How is it that we can be governed by such contradictory and unacceptable impulses? I have felt the thrill of thievery and savored the tastes of malice and successful deceit.

Among the first of our moral codes was the Ten Commandments. Do we imagine they were written in stone to reprimand the rare individual who ran afoul of them? It's more likely that Moses recognized that these were ten persistently present weaknesses which plagued people's inner lives every day. I don't actually kill people; I just imagine killing them with regularity. And that's the commandment I observe the most admirably; things go downhill from there for the remaining nine.

I like to think I'm better than that. At times there is better in me; at times there is not. I don't want anyone to know this about me. With no one keeping watch, malevolent impulses arise veiled but unchecked throughout my day, sabotaging good intentions, while I pray no one notices they are there.

Most run away from fear, but some people turn toward it and act well despite its presence. We call them heroes. That's why everyone admires firefighters. That's why the Buddha smiles.

People often ask me, "What is the difference between a good carpenter and a great one?" It's a difficult question to answer because I've never met a great carpenter; I don't know that the systems of training are in place that might produce one today. Five minutes in Versailles will convince any craftsperson that there is a long, long way to go. I've gotten into the habit of demoting the question to "What is the difference between an okay carpenter and a good one?"

My answer has always been: "Fear."

At all the disparate places I've worked, I have been surprised that, when the opportunity arises to build something out of the ordinary, few people step forward to volunteer. The most interesting and challenging projects were always available to anyone who was willing to say, "I'll do it."

What was everyone else thinking?

"I've never done it before"?

"I'll get in trouble if I screw up"?

"Let someone else fall on their face"?

All three are fair enough arguments. There are plenty of things I shouldn't try to do, at least not in public. Surgery and the mambo readily spring to mind. I imagine that my duodenal butterflies would stir if I undertook either. Medicine and Latin dance are pretty far outside my experience and training, but even those I might venture if the opportunity presented itself and the rewards were desirable enough.

By comparison, taking on a challenging project isn't worth hesitating over. I want to learn more, to become more proficient, to make things that I think are astonishing. If that means enduring days or weeks of self-doubt and intestinal discomfort, then that is the price I must pay. I face real consequences when I screw something up. But unlike surgeons and professional dancers, I have had the chance to fix the majority of the mistakes I've made in carpentry. I have to live with the imper-

fection of my work. Imperfections are the first thing that come
to mind every time I remember a past project, but most of the
fun and satisfaction I've had in my career has come from finding solutions to problems nearly everyone around me turned
away from.

For the last thirty years, every job I've built has included an
assembly no one has ever tried to build before. By now, I've
developed a reputation for saying yes to these things. I've built
hundreds of devices and structures that people live with every
day, all of which I had never previously attempted. On the Blue
Tape job, I counted fifty.

In manufacturing, companies usually take years to work
through a product design. They make series of prototypes to test
and perfect their creations. In high-end building, we make one.
It's probably the most idiotic business model ever devised. It's a
wonder I'm not in the poorhouse, sued out of existence by
some client whose rotating sushi bar airfoil screen didn't spin to
their satisfaction. Idiotic or not, I can't seem to stop doing it; it's
too much fun.

Construction used to keep me up nights. I'm not aware of it,
but I've been told that of late I work in my sleep, giving audible
instructions to my co-workers and sweating through the night's
mishaps. Fear is a constant companion underlying much of what
I do. But because we are old pals, I now hand the narrative over
to Fear, who has cleverly combined dreaded elements from my
childhood with nagging anxieties of today. Together, we present
this report card evaluating some of my more notable inventions.
I can hear the butterflies approaching. . . .

LIFETIME REPORT
SEPULCHER OF THE SCARED HEART DAY SCHOOL

A—Superior   B—High   C—Average   D—Poor   F—Falling

GROWTH IN SKILLS AND PROJECTS

ASSIGNMENT	Concept	Execution	Appearance	Performance	Final
MATERIALS					
Site Cast Concrete Hearth and Cantilevered Bench	B	B	C	B	
Ultra-thin Q-Deck and Plywood 4" Floor Assembly	B	B	B	B	
7 Story Sculptural Wood and Painted Staircase	A	B	B	B	
Flexible Cast Molding for Vaulted Powder Room	C	C	D	C	
Glass and Stainless Catwalk w/ Cast Bronze Rail Details	A	B	A	A-	
Sculptural Facetted Volume w/ Sintered Glass Stairs	A	A	A	A	
Curved Sculptural Epoxy Composite Staircase	B	A	B	B+	
Perforated Steel Staircase w/ curved wooden guardrail	C	B	A	B	
METHODS					
Hand Bending and Shaping Brass Railing Part	D	B	B	B	
Cantilevered Room-Dividing Pocket Door	D	C	C	D	
Modifying Gas Fireplace Burner to Mold Flame Shape	D	C	F	D	
Polyurethane Cast Oculus Windows w/ Integrated AC	A	B	B	B+	
Truss Support to Replace Foundation under House	B	D	NA	A	B+
GEOMETRY					
Elliptical Domed Entry w/Fanlights and Pocket Doors	A	B	A	A	A-
Art Nouveau Limestone Staircase w/Forged Railing	A	A	A	A	A
Intersecting Elliptical and Radial Ceiling Vaults	C	B	C	B	B-
Sculpted Entry Ceiling Above Radiused Staircase	A	D	A	A	A-
TECHNOLOGY					
Hinge for Rotating Plasma Monitor Wall	A	B	D	C	B
Robotic Arm for Control Panel in Porsche Garage	A	D	C	C	B
"DaVinci Machine" Art Light on Curved Track	A	B	A	A	B+
Housing for Creston Control Panel/Tilt for Ease of Use	B	A	A	P	C

A-Always   B-Usually   C-Sometimes   D-Seldom

GROWTH IN HABITS AND ATTITUDES

Decade	1	2	3	4
SOCIAL HABITS				
Respects Authority	D	C	C	D
Takes Correction Well	D	C	B	D
Respects Property of Others	B	B	B	B
WORK HABITS				
Does Required Assignments	C	B	B	A-
Follows Orders Accurately	C	C	B	B
Starts and Finishes Work Promptly	D	C	B	B
Keeps Work Area and Materials Neat	D	B	C	B
Responds Well to Questioning	D	C	C	C
Wise Use of Leisure Time	D	C	C	C
PERSONAL QUALITIES				
Accepts Responsibility	C	C	C	C
Exercises Self-Control	D	C	C	C
Personal Appearance	D	C	C	C
Manners—Courtesy—Politeness	B	B	D	B
*Prayer at Home	D	C	C	D
*Reception of Sacraments	D	D	A	D
*Thrift—Wise Use of Money	D	D	D	D
Days Absent	237	84	9	17
Times Tardy	212	45	12	13

Parent cooperation is essential to educational growth.

Please sign and date this Report Card after review with your child.

signature   *date*

Property of Sepulcher of the Scared Heart Day School

There, that's done.

I've only taken one construction job solely for the money. It wasn't a terrible job; no one wanted me to violate whatever principles I might hold; there just wasn't much about the job that interested me. The man who hired me offered 30 percent more than I was making; he talked about having me join as a partner someday. With three boys at home and college only a few years off, it seemed like the prudent thing to do.

Prudence is a feeble reason for doing anything. I suppose security has its place; constantly working at the edge of my abilities is stressful. But, looking back, every time I tried to sidestep fear for security, I wound up unhappy. The new job was miserable.

Ostensibly, at least in name, this was a promotion. My business cards arrived with PROJECT MANAGER printed on them; I tucked them into a drawer and never touched them again. In practice, this meant that most of my building knowledge was tossed out the window so that I could spend my days soliciting bids and writing contracts, entering endless heaps of data into an overpriced computer program mockingly named Timberline, as if anyone glued to a computer monitor would ever lift their gaze *that* high again.

The new position wasn't a complete waste of time; it forced me to learn the side of the business that never interested me: the business side of the business. I learned how to concoct proposals, how to compose invoices in the ways most likely to ensure payment, and the value of preemptive communication, especially when things started going wrong. I'm glad I learned the position, but I never want to spend my days doing it again. There is an art to this stuff, and there are people who love it; they are the ones who should have these jobs.

After enduring project management for four years, I was offered a 10 percent stake in the business. Perhaps it was only 5 percent—it doesn't matter; I didn't even consider it. Things in this company were clearly going wrong. The owner was a lean smooth talker, with a Caesar haircut and a lippy Brooklynish accent that seemed put-on. He was far more interested in construction as a cash-absorbing machine than as an enterprise. His real interest, other than wishing to appear rich, was the restaurant business in which his family was involved in ways I never hope to know. For a few months leading up to the partnership offer, things on the construction side were being neglected in favor of the restaurant side. Subcontractors weren't getting paid; new projects weren't getting under way; none of the project managers were able to buy fixtures or hardware for their jobs.

Smooth Boss would call regular meetings filled with assurances that cash flow was being addressed and money was coming in. But the man, rail thin when I first met him, was now skeletal; his movements had become angular and terse. Whatever stresses he was suffering, and shell games he was running, they weren't worth it; they were eating him alive. No one believed a word he said; most had already begun looking for a way out. It appeared that my "partnership" was an invitation to become part owner of a mountain of debt.

Partnerships and impending bankruptcy aside, so long as I still worked for the outfit, there were daily tasks that needed attending to. I was responsible for running the renovation of a project in one of the few remaining mansions on Fifth Avenue. It had been the city residence of the same coal baron who built The Elms in Newport, Rhode Island. He was long gone, but Gilded Age trappings haunted the place. A newfangled baron of some sort was fixing it up for his wife and family. I never met him. Apparently, he had become aware of his imminently impending death. He was holed up in Paris awaiting the end, and was seeing to it that those he left behind were properly cared for. He entrusted the renovation to a remote business manager, who in turn entrusted the day-to-day operations to a local supervisor, who was compensated in part with an apartment on the premises.

This supervisor was the only representative with whom I had direct contact. He was a born and bred New York guy, fiftyish, from Queens judging by his accent. He had a light smattering of building experience and alluded regularly to his "union days," which he assumed lent him credibility. The overarching impressions he made were that he was way out of his depth and that he was petrified of his superiors.

The renovation was an opulent one. Silks, semi-precious stone, gold plating—it was all spread about the place in the impressive yet unremarkable way in which so many New York apartments are decorated by adherents of what we in the trades crudely call the "Big Dick" style.

In a project such as this, procurement and purchasing can be just as complex as the work itself. A half-million-dollar container of semi-precious Jaspe du Var stone slabs was ordered from a French quarry for the master bathroom suite; the ship encountered a storm in the North Atlantic, and a container of costly rubble arrived on our shores. Who pays for which calamities became an urgent question.

The super from Queens was handed the sword of Solomon and ordered to sort it out. He wielded it with the recklessness of Oedipus and to similar effect. It quickly became impossible to buy or do anything. He requested three bids for everything. I was forced to make weekly explanations of accounting realities that included nuggets of truth such as "Butler Hardware is the only company that makes Butler Hardware. I am unable to secure alternate pricing for this line item." Tiresome is too weak a word to describe the process; loathsome is closer to the mark.

It was over the issue of hardware that discussions deteriorated entirely. Architects customarily provide lengthy hardware spreadsheets for the operative aspects of a project: door hinges, knobs, thumb turns, locksets, those sorts of things. When a project is sent out to contractors for initial pricing, it is rare for the decorative hardware, like drawer pulls and cabinet latches, to have already been chosen. Astute contractors routinely exclude it from their bids. Smooth Boss was so distracted by his restaurant ventures that he neglected to exclude the decorative hardware from this project's estimate.

Queens Super put me in an untenable position, insisting that I purchase tens of thousands of dollars' worth of cabinet knobs and pulls, some of them custom plated in 24 karat gold, with no budget. Because for a while it had been nearly impossible to pry checks out of the accounting department for *any* purchases, it seemed like a good time to set up a meeting between Queens Super and Smooth Boss so they could hash this out.

The result of this meeting was that we would rely on a time-honored and pitfall-laden convention of general contracting. Smooth Boss agreed that we had "bought the drawings," meaning if something could be found in the stacks of pages predating the contract signing, he would be willing to provide it, no matter the cost. This was his contract and his call. A meeting was set up between Queens Super and me to determine which decorative hardware was in the drawings and which was not.

On the appointed day, I arrived onsite with a full-size, freshly made set of blueprints dated just a few months before the contract's signing date. I also brought a copy of the current prints so that I could demonstrate that no other drawings had been issued between those two dates. Queens Super met me in the sitting room at a folding table set out for our purpose. He was visibly agitated; this was a responsibility he did not want. Perhaps he had been threatened with the prospect of purchasing whatever hardware wasn't found from his own paycheck. I don't know; we didn't discuss it. In fact, we didn't discuss anything that day.

I leafed through several handfuls of generalized layout drawings, noting that no decorative hardware was depicted in their pages. Finally, I came to the section labeled "Millwork Detail Drawings." On the first page was a set of drawers for the dressing room. "There are several drawer pulls shown here." I circled them with a highlighter. "We're responsible for these. Okay?" Queens Super nodded with a grunt. I turned the page. It showed a sideboard that was to be built into a long niche in the dining room wall. "I've looked this one over several times; I don't see any hardware drawn or specified on this page." No noises escaped from him, not even a grunt this time. "Do you see any hardware on this page?" Still no grunt . . . nothing.

I took a few breaths, hoping it would help me think. He stood motionless, his arms pressed to his sides, his eyes sightlessly fixed on the blueprints. "Okay, I'm looking at this page of drawings. As I look around it, I can't find anything that looks like hardware. Do you see anything that looks like hardware?" Jesus, *still* nothing. I felt the grip of oncoming agitation. "If you see anything that looks like hardware, could you please point it out to me so that I can see it, because I'm missing it."

I have no idea what was going through his mind, but none of it was coming out of his mouth. This was far too unreal. I lost it. "Would you say something? 'Cause I think I'm standing in a room talking to you about hardware, but I might be out of my fucking mind, I might be in Canada or asleep on the beach, so if you would say something, or move your arms around a bit, it would make me feel better about this." He just stood there. He didn't even look at me. He stared at the drawings and stood there. I rolled up the blueprints and walked out.

Later that week I learned that Smooth Boss had torpedoed his construction company and there wouldn't be any more paychecks. He embarked on a new career, opening branches of his most popular little eatery all over the world. He passed out special cards to all his valued construction employees, most of whom he owed weeks of unpaid salary, allowing them to jump the take-out line at the swanky boîte. We gave them wrapped in twenty-dollar bills to homeless people and encouraged them to buy whatever they wanted. I don't know what retribution his jilted clients cooked up for him, or if he survived it. I never did order any hardware.

FAILURE

Always get out before the authorities arrive.

I've done everything wrong at least once. I don't remember clearly, but I'm guessing this pattern picked up at the outset of things. I've seen three sons born, and not one of them got a hold of a nipple successfully on the first try. Since I'm no better than they, I will assume that failure for me started there.

Somehow a great number of people have been conditioned to view failure in a less than positive light. This is easy to prove. Think back to your childhood as you read the following sentence:

"Failure, Failure, Failure, Failure, Failure, Failure, Failure, Failure, FAILURE, FAILURE, FAILURE, FAILURE!"

Take stock of your emotional state: Positive or negative?

It's a long walk from "You can do it, little guy!" to "You fricking idiot, you just cost me twenty thousand dollars!" Both are forgivable responses to a given circumstance.

Lest anyone think I am exempt from shame, I record this episode as a means of reminding myself otherwise.

Young parents are torn every which way by life. Nascent careers, newly acquired spouses, sick kids and healthy ones, and not enough experience to be good with any of it make young parents an unreliable bunch. Among themselves, they are quick to swap stories and form alliances, like freshly pledged members of an ancient fraternity; they have a natural compassion for one another, and many who don't share in their predicament do share in forgiving their pitiable state.

With three preschool-aged boys at home, I was hired to build a deck and hot tub enclosure in Pound Ridge, one of New York's tonier northern suburbs. I was offered the project by a young architect who had done previous work for the owners. She was the single mother of a young daughter, knew I needed work, and supposed I had the skills to build the project.

I was indeed fairly skilled by then, with seventeen years under my belt; what I lacked was the organizational capacity to contract the work. She showed me her designs for a modern extension to the house, which had been designed by another modernist architect of some note. The house was a hodgepodge of intersecting rectangular volumes and her extension reflected it in a suitably reverential fashion. Without visiting the property, I dashed off an estimate, as optimistic in its confidence of my carpentry skills as it was unaware of the obstacles I would face working seventy-five miles from home in an outdoor environment with which I was unfamiliar. The price was a bargain; I came nicely recommended; we were in business.

The first day brought the first signs of trouble. Driving through New York during rush hour, it took two and a half hours to reach the place. When I arrived, I inspected the existing house and realized that it was entirely clad in a siding size that no lumber company carried. A local boutique building center carried a size a few inches wider, in a lesser grade, at a price far above what I had predicted. I winced and placed an order. The next three days were spent re-sawing, routing, and sanding my stack of lumber until it matched that of the house. Three days of my two-week schedule were gone, and I had not even begun to build.

I called a friend, another young father who painted apartments in the city for a living, and asked if he could help. Luckily, he was as desperate as I was, so we struck a bargain. I offered him more than my budget would bear, but I comforted myself with the well-worn contractor's delusion that the "profit margin" would somehow cover the difference. Mimicking companies I had worked for, I had added 20 percent to the prices of the various tasks required by the project, calling it "Profit and Overhead," while somewhere inside me a financially inept voice cooed, "Oooooo, extra money!" Craftsmen often make up for in optimism what they lack in business sense.

The next morning, I picked him up in Brooklyn, adding a half hour to my trip, and we set off for work. Things went better. At the end of the second week, we had finished most of the carpentry, so we pried a sample of the original siding from the house and went to the paint store to buy a few matching gallons. The paint seller performed a "spectral analysis" of our sample and happily mixed up two cans of it, at three times the cost I was expecting. Two days later we had primed and painted the entire deck. The color was nothing like that of the house. No amount of yammering about "different light" or "it will age to match" could mask the discrepancy; we didn't even try.

We painted a scrap board with the new paint, pried the old board off the house again, and headed back to the paint store. The spectral analyst agreed that there was a noticeable difference between the two boards. He offered to "bump the tint" for a better match. I wasn't bristling with options, so I let him give it a try. Two cans and one day later, our addition matched no better than the first time.

One year of my schooling was spent in a freshman foundation year at a New York art school. It had all the rigor of summer camp, but I did have a color theory teacher who really knew her stuff. I gave up on the spectral analyst and did my best to remember everything she had taught me.

This house was gray. Grays can be especially hard to match, because mostly what one perceives of them is the value—lightness or darkness, and the subtle undertones that can be from any color on the wheel. I went back to the paint store and bought a middle gray slightly lighter than the house, with the most neutral undertones I could find. To move the undertones in the direction of the original house, I purchased a rack of tints featuring every color on the wheel, most of which I knew I wouldn't need. My painter friend and I mixed up sample after sample and painted each on a primed board, knowing the color would shift as it dried. A day later, we had our formula. Back to the paint store we went and in another two days we were done, a dead match.

This is not success. Both the material budget and the work schedule were double what I had assumed. After paying my friend

I made about nine hundred dollars in four weeks. I'm being kind to myself; I had spent a lot on gas. Still, the owners were thrilled; the addition matched the house nicely in quality, character, and color. Apparently, none of their previous contractors had been able to pull that off, so they asked if we could add a few windows to one side of the house and restore the siding. I was happy at the idea of redeeming my "profit margin," so I agreed.

This time, I prepared my price more carefully, looked it over critically, and doubled it.

The windows of this house weren't commercially made; they had been designed to match the minimal house's minimal details. Really, they were simple wooden frames with a single piece of glass. This was before I had my own workshop, so I arranged with a cabinetmaker friend of mine to rent his shop for two weekends. My helper and I made the frames, painted them with our secret formula, milled another stack of siding, loaded the truck, and drove back up north to install everything. A local scaffolding company delivered several sections of pipe frames. We set it all up and got to work. One of the upper windows was meant to double the size of an existing one. We carefully tore that out and cut a new opening big enough for ours. The first two days went well; we moved along one window at a time and had all the upper windows in place, more or less on schedule, after a week.

Monday of the next week, we returned to work. It was raining miserably. After a few hours we gave up and made the drive home. All week long the rain continued: We couldn't work; the schedule and budget were slipping away again. Another Monday came, and we tried to redouble our efforts, deciding to demolish all the remaining siding on that level in one go, hoping it might increase our efficiency. We were only a few boards in when it became clear that we were in trouble. Behind the boards of the lower level the plywood sheathing was wet and rotting. We peeled away the plywood and found that many of the studs had rotted through and through. We surmised that the entire side of the house was being held up by rotten studs, withering plywood, and siding that wasn't capable of supporting anything. I was not a house carpenter; I was a New York City interior renovation carpenter with scant experience in suburban home-building. My friend and I were now in over our heads.

We decided to investigate further, hoping we could gauge just how bad the situation was. We pulled off an entire piece of sheathing. The wall behind was teeming with termites, healthy over-fed swarming colonies of them. For the rest of the day, we pulled siding and plywood off of the house. Everywhere we probed, happy clans of insects were busily eating the house we were supposed to be fixing. I had no idea what to do. The day finally came to an end. My friend and I drove home. Neither of us had a way of resolving the situation. I was broke; he was tired; our wives and kids had barely seen us in more than a month; we had reached our end. I dropped him off at home, thanked him for his help, and told him that was it.

The next morning, I drove back to the jobsite, dismantled the scaffold, and called the company to come pick it up. I drove off like a coward.

I was certainly afraid to face the owners. I had seen more than one bearer of bad news eviscerated by a wealthy client. More-over, I hadn't even made carpenter's wages for my fifteen-hour days. I didn't have the carpentry or business skills to fix their problem. I was failing as a provider and absent as a father.

Broke and defeated, I never called the homeowners, and they never called me. I have no idea what became of their house. Hopefully, a real suburban contractor with actual solutions got them out of the mess. The fix couldn't have come cheap.

To this day, I won't give a set price for a project. I mistrust both my financial acumen and my optimistic outlook. After forty years, everything I build is new to me in some way and will have its surprises. I charge an hourly rate; I work hard and fast, and I try mightily to get things right. My carpentry skills are far beyond what they were in those days, but my business skills have only developed enough for me to know that I need to be paid for every hour I work. If there's a serious problem, these days I have the guts to talk about it; I'd rather not pile up any more mud on the cowardly side of the fence. Clients can fire me whenever they like. No client ever has.

THE MAGICAL STRETCHER

Every once in a while, someone gets famous. Life has no shortage of monstrous curses, but this has got to be in the top three. Fame has its challengers for the Curse Crown: being born too beautiful or too rich to have to try at anything rival fame for devilry. All three are mistaken for blessings. Monsters masquerading as angels are the most infernal kind.

My work has brought me into the spheres—well, more accurately, the homes—of a handful of famous people, and another armload of semi-famous folks. Most of them are dead now, or close enough. Things often didn't end prettily for them.

Fame, wealth, and beauty are the primary standards of success these days. They are also the primary objects of envy. People imagine that all three bring the contentment many feel they lack in their lives. In the years I've built for people who possess any of the three, I have seen little that confirms that belief. My clients outstrip my co-workers in wealth, social position, influence, and resources, but they also outstrip them in categories they'd rather I didn't mention: chemical dependence, fractured relationships, impatience, irritability, dilettantism, and malice. I've met precious few exceptions.

   Success and failure are poor lenses through which to view a life. What matters most?

1. Others perceive us as successful.
2. We think ourselves more successful than those around us.
3. We have done what we set out to do.

   Each of us has to live with ourselves; other people have to live with us, too. I think contentment, fulfillment, and decency are better measures of the quality of life one leads. Each of us must satisfy ourselves that our life has been worth the time we've spent living it.

Of the many people for whom I've worked who are considered highly successful in their fields, there are only three who have lived public enough lives that people know their names: David Bowie, Robin Williams, and Woody Allen. My time with each of them was short, so there is nothing to ogle at here, but all three encounters were instructive in their way.

   Of these, Mr. Bowie brings the fondest recollections. I've played music my entire life, longer than I have been a carpenter. The idea of working on his house was exciting to me. I'd get to meet an actual hero of mine; I don't have many. My firm was hired to install several rooms of cabinetry in an apartment he and his wife bought almost thirty years ago. One morning, while eating my egg sandwich, I heard a rustle behind me and a gently offered "Good morning." I turned to answer. Standing across the room was a man I had admired since I was a teen. He was slender, elegant, kind faced, and alert. "When do you think this place will be ready?" he asked. "Do you want me to tell you what I'm supposed to say, or what I really think?" "Hmm, that's all the answer I need. Enjoy your breakfast!" He smiled and we never spoke again. I will always like him.

I also felt bad for him. His apartment windows were fitted with two-inch-thick ballistic glass. An inner room was lined throughout with three layers of half-inch Kevlar, enough to stop a bazooka. The room was equipped with an air filtration system and a radiophone like ships used at the time. He had a lifelong fear that someone would shoot him dead someday. John Lennon's murder had only exacerbated his terror. He wasn't some highborn VIP or celebrity's kid: His mother worked as a waitress and his father for a children's charity.

He reinvented the words, melodies, arrangements, and images of popular music. He shouldn't have had to live in fear. Perhaps fame was a curse to him. It didn't keep him from treating a carpenter he didn't know with two minutes of decency.

A few years later, I worked on the apartment Robin Williams fixed up after his divorce and remarriage. It was a modestly swanky place just above the tree line of Central Park. Again, I had stopped to eat, this time a hamburger deluxe. I was staring out the window at the tops of the trees. The wind blew mildly, and the leaves made swirling light and dark patterns as when it blows through an animal's fur. It was fascinating to watch, a different way of seeing trees, akin to the feeling one gets when unhearable sounds are amplified to audibility. Mr. Williams appeared over my right shoulder and began talking, not really to me, just talking, about the park and the weather and so on, in monologue form. It was long ago, and the only thing I actually remember him saying was "Enjoy your hamburger!" At least celebrities share a concern that I am well fed.

I felt bad for him, too. His mother was beautiful. He learned to be funny to win her attention. Soon enough he won the entire country's attention, mine included. Our brief interaction was like the one-way interaction of a performer with his audience. Perhaps that was the quality of attention he got from most people. It didn't help him.

Woody Allen's apartment sat atop a nondescript Fifth Avenue prewar. We weren't there long, just installing a few things, hanging doors and such. I walked through the kitchen once where he was standing. I said my "Good morning" as people do. He turned slightly to the side and remained that way until I passed through. I understand his distraction: He was being pilloried by all of society at the time; he likely trusted almost no one. In his shoes, I probably wouldn't have looked at me either.

These limited brushes with fame took place in the span of a few years, little more than a decade into my career. All three left me unsettled. For most of my life, fame has confined itself to a few public spheres that others occupied. As a carpenter, it is not something I ever thought I would have to worry about.

Just a few months after working at Woody Allen's house, with nothing to do one evening, I decided to watch a television show that had been recommended to me. Much to all my colleagues' surprise, a carpenter had become famous, and he had been engaged to star in a show demonstrating how things are made. I thought I would take a look.

This evening's project was a common stretcher table of traditional design. It's a simple idea:

1. A slab tabletop is made.
2. Four legs are arranged in a rectangle.
3. A horizontal "skirt" of sturdy boards is tenoned to the upper end of the legs to hold them vertical.
4. A "stretcher" of thinner wooden pieces is arranged near the legs' bottom to keep them in place and offer extra support.
5. The portion consisting of the legs, skirt, stretcher is assembled.
6. The slab tabletop is attached to the leg assembly, completing the table.

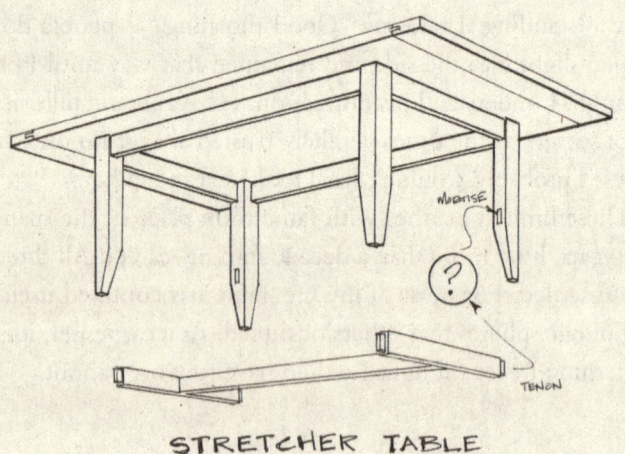

MORTISE

?

TENON

STRETCHER TABLE

I watched the work progress with interest. The man had nice tools—maybe he had sponsors? No, they can't be sponsors; the labels are obscured. My inner critic followed as he worked through each step. The tabletop slab was neatly finished, with breadboard ends, a nice touch. Then he put together the leg assembly. He slid the skirt's tenons into the legs' mortises and glued it all up.

"What the fuck, you dipshit!" I shouted at the screen. "You

forgot the stretchers! How the hell are you going to get those tenons into the legs without prying the whole thing apart?" Now I was really interested. I've made so many dumb mistakes just like this in my career. Failure is at the root of almost every reliable technique I use. There's drama to be found in anything, and this is where drama is at its richest in the building world: when things go wrong. I sat up, finished my coffee, and waited out the commercial break, eager to see him make good on his error.

The next shot was of the completed table in front of a tasteful gray background. The unseen carpenter's voice came from the TV: "All that's left to do is glue the stretchers in place, attach the top, and your table is done."

"What! What the fuck! You fraud! Chickenshit bullshit! How'd you get the stretchers in? How is anyone supposed to learn anything from this crap?" Was it too much to ask that this veteran carpenter admit that he still made mistakes? Did the show's producers think that owning up to an error ruined his credibility as a craftsman? He screwed up, and fixing it was going to be messy. *Everyone* screws up, and fixing it is almost always messy. A show about how to make a stretcher table is marginally interesting at best. A show about facing one's flaws would be downright fascinating.

I never watched the show again. I had learned the only lesson it had to teach:

Fame does terrible things to people.

For a year in the early nineties, I shuttled between the summer and winter homes of a mergers and acquisitions man. Our schedules were so completely opposite that we only crossed paths once, and that was at two in the morning when we both happened to be working at his Florida property. I was twelve feet in the air installing a fourteen-piece hand-carved crown molding in the entertainment room. And he—well, I don't know, we weren't allowed into that wing of the house—I imagined him working in an elliptical, gilt-lined office with a hand-painted choir of angels on the ceiling above, egging him on to bigger and more ruthless deals. When he wandered into the room where I was working, he clutched an outsize unlit cigar, wearing a striped silk robe and what appeared to be little else. He cut a sort of "Hugh Hefner meets Barry Goldwater" figure. He lingered and admired our work, and I was likewise impressed that he was still at his, even at this late hour, despite the fact that he had reached the station where dinner for life is all but assured.

The weather set our schedules. Just as it became reliably warm north of New York City, M&A Boss would relocate to his home there, and we would move our operation to Florida, where things were just beginning to swelter. Then, as Florida settled into a balmy-breezed agreeability, we would switch places, landing us in the North's drab gray wash. He relayed to us only one mission through his point man Steve: Finish everything before the seasons change and switching time comes.

To this end, Steve reconfigured everything about how projects are conventionally done. He hired the same highly reputable New York City architecture firm for renovations at both residences. This outfit was required to draw everything in the most minute detail. Thick sheaves of hand-drawn blueprints arrived depicting each room scheduled for refurbishing. Steve reviewed every mark on them, made lists of anything lacking, and required revisions in a few days' time. Once he was satisfied that the drawings were complete, the architect was dismissed and told he need not come around again.

Tradesmen were assembled based on three criteria: competence, reliability, and endurance. I worked as a subcontractor/installer for a cabinet company that made the most beautiful woodwork I have seen before or since. On the date Steve had meticulously scheduled, foremen from a dozen or so outfits would arrive at the home in question. On this occasion, a line of vans formed in the driveway of one of Palm Beach's largest ocean-side mansions. We had each been sent drawings pertinent to our trade weeks before and were expected to have studied them well. My job was to measure every existing component in the rooms that were to be renovated and notate the drawings so that everything the shop made would be perfectly tailored to the space. Once my survey was complete, I was expected to make a list of any questions I had; usually there were dozens.

In a single two-hour meeting with Steve, I would have all my answers. This one- or two-day process has never taken anything less than weeks with any architect or designer I have worked with. On some projects, it has taken the better part of a year. (On most of my largest projects, this process never reached completion; I would simply give up hoping for answers, draw assemblies myself, and send them to the design team for approval. Many of these assemblies can be found uncredited on those same architects' and designers' websites.)

I'd work on the drawings with my red pen for another day or two, until I was sure that every detail was clearly shown and every dimension was correct. One copy was sent to my employer's workshop for fabrication, and I kept one so that when my installation window came, I could refresh myself on the job's details and be prepared to begin work as soon as my crew and I returned. In forty years of building, this was the most efficient system I've ever seen. No hemming and hawing in "progress meetings," no "We will have to revisit that" from indecisive design professionals, just ask, answer, build, an unheard-of trifecta in our industry.

This was the wonderful side of these projects. We built grand things, in palatial spaces, in record time. Many of Newport's mansions were built with the same startling speed and effi-ciency.

By the time we built these rooms, minimalism's spartan ex-panses had become high design's standard of elegance. But minimalism has a nemesis, and it is found here. We produced excess at its best. Rare woods, rich handwrought detailing, gold-plated hardware, seamless finishes applied in situ that glowed deeply in any light. Minimalism can argue its claim to aesthetic purity, but like any well-run bacchanal, this stuff had a lock on fun.

Steve didn't care how we behaved, how we dressed, what music we played, or what antics we got up to as long as we pro-duced. The guys on my crew were there precisely because they loved to build beautiful things, and they loved to do it fast. We even had a motto that we would intone in an Orson Welles–like voice: "Quality is not a speed." So Steve liked us, we liked him, and we could get away with almost anything.

Each morning we arrived at the steel service gate in a plain white rental van. Jeff, my most skilled co-worker, was a fidgety Mainer who enjoyed getting under people's skin. The squawk box would crackle as we pulled up: "How can I help you?" Every morning, Jeff's answer was different, but his favorite vein went something like this: "Ahmed! Ahmed! It's me, Salim! Open the gate! I have the fertilizer!" The shrill staticky answer would come back: "This is Security. This shit has to stop, NOW!" Jeff would laugh himself silly and come up with a new variation the next morning.

For Jeff, second best was to chat with the young bodyguards, mostly ex-military types with no-nonsense attitudes and the physiques to match. "Lemme ask you," he'd prod in his Italo-Mainer accent, "how much do they pay you to do this job?" No answer. "I mean ballpark, fifteen, twenty dollars an hour?" Angry silence. "So, dude, really, if it came down to it, would you take a bullet for this motherfucker for twenty dollars an hour?" Jeff wasn't very well liked around the place.

One morning, when we had just gotten going, Steve came in with a midsize man with dark hair and eyes, and a face that didn't move, even when he spoke. It seemed like his whole body would remain completely motionless. Steve introduced him as the head of Security. There was something about him—even Jeff understood it to mean that this man was not to be toyed with. Somehow, without moving a muscle, he exhaled words in an Israeli accent: "I won't have you talking to my men anymore. Understood?" He didn't really look at us. It seemed as if he were looking deep into a past of dark rooms holding solitary, weeping men. "Uh, okay," Jeff murmured. That was it for the fun.

Thus began the less wonderful side of this work. The summer was an endless parade of sixteen-hour days, with Sundays off for sleeping. Florida in late summer is an unholy place. We would work in boxer shorts and boots, wishing a wisp of air-conditioning could find its way to our wing of the house.

We completed the one-thousand-square-foot pilastered and paneled English brown oak entertainment room and moved on to a pearwood dressing room. Pear has always been my favorite wood, soft and smooth, with an iridescent shimmer when it's shaped and finished well. The dressing room had a barrel-vaulted ceiling, slender turned columns at the uprights, with gold and etched mirror accents between the hanging bays. It was the most perfect room I have ever made, lovely everywhere you looked, understated and richly tasteful in the interplay of colors,

materials, and shapes. I guarded it angrily whenever an un-
trusted tradesman wandered in hoping to find a place to stash his
tool bag. No scratches in *my* dressing room!

At last, the summer ended and our work was complete. M&A
Boss was heading to Florida that week and we were headed for
his New York home to redo several bedrooms. A weekend
passed; Steve called to let us know he would be joining us the
next day, and to tell us the owner hated the dressing room. He
had Steve rip it out.

Like the best theater, it lasted a week and was gone forever.
The only reason we were ever given was that his valet thought
it should have been larger. He felt that M&A Boss's suits didn't
hang in it loosely enough.

The New York house was smaller than the Florida one, seventeen thousand square feet to Florida's forty-two thousand. It
was built in the English vernacular style, half-timbered in hick-
ory, with antique brick infill and a custom glazed terra-cotta
roof. It could have been called quaint, except that quaint doesn't
come in that size. The rooms we were working on were second-
ary ones, nothing so grand as the millwork we did in Florida.
When Steve arrived that week, masons were already staking the
foundations for his main concern this winter—a new two-story
garage styled to match the house in every detail. His deadline
was April 1.

The garage was an unimaginable undertaking. Sourcing the
materials alone would have been the job of one man for months.
Steve took responsibility for everything.

Jeff and I barely slept; it seemed like Steve never did; he rarely
even sat. By January, I was exhausted. Our work on the bedrooms was completed on time. Jeff and I packed our tools and
were preparing to leave when Steve called me in to his office.
The boss had just bought a one-hundred-thousand-square-foot

villa in the south of France, and Steve wanted me to go with him to run the millwork installation. I told him I'd think about it.

I never did. I had been working ninety-six hours a week for a year; I had barely met my two-year-old son; and hearing about the dressing room had plucked the last shards of joy from my work. I was done. Even working enough hours each week to earn nearly three times what was already a generous salary couldn't provide the motivation I needed to take his offer.

Steve confided that he had learned a good deal about investing from M&A Boss. He planned on finishing the French villa and retiring upstate. I wished him luck and shook his hand goodbye.

By early March, I had found a new cabinet shop to install for, working regular New York City building hours. It was far less money but a huge relief, almost like a vacation. To my surprise, a few weeks later, a call came from Steve; he asked if I would drive up and help hang some doors. I was well rested and grateful for the extra work, so I agreed to meet him on the following Saturday, so I could work the entire weekend. I left my house at five in the morning and made it to the estate at seven.

The garage that had stood nearly completed when I'd last been there was completely gone. There was nothing but a mound of freshly spread topsoil in its place. I found Steve in his office sitting at his desk. His eyes were ringed in red, hollow, and dilated. He was crying and unable to care that I saw. I asked him what had happened, what the matter was. "The boss got here last Friday. We finished it; we fucking finished the garage. He called me from the helicopter and told me he hated it. He's coming back today, and he wanted it gone. The guys have been bulldozing it all week; the last truck just left. I don't know what I'm going to do. I'm done. I'm fucking done. I can't do it. I'm done." He stared at me with the look of a man who has nothing left.

Several years later, M&A Boss's upstate New York house burned to the ground. By some stroke of luck, in a house with a full-time staff, no one was there when the alarm went off.

I am accustomed to the idea that everything I build will be destroyed. Most of what I build probably lasts around ten years, until the next co-op buyer comes along and deems it outdated. Whenever I begin a new renovation, I try to spend a few days salvaging anything that can be reused. Like me, most every carpenter I know has a house with a few absurdly opulent features that have been scavenged. It doesn't pain me too much that the things I've made are gone; I haven't had to witness their destruction.

Most dedicated craftspeople make things as skillfully as they can. When, through our efforts, we are able to make something truly beautiful, it is a source of real pride, but pride with an element of detachment. The things I've made have never been mine. My work is to make things for other people. Some of those people are lovely; many are indifferent; a few of them are the most destructive people on the planet. It can be a difficult world to live in; a billionaire capriciously waves his hand and the most perfect room I have ever built is torn to pieces. I had to quit to recover. He waves his hand again and a man who has accomplished something nearly impossible is forced to destroy his own accomplishment. I don't know if anyone recovers from that.

+ + + +

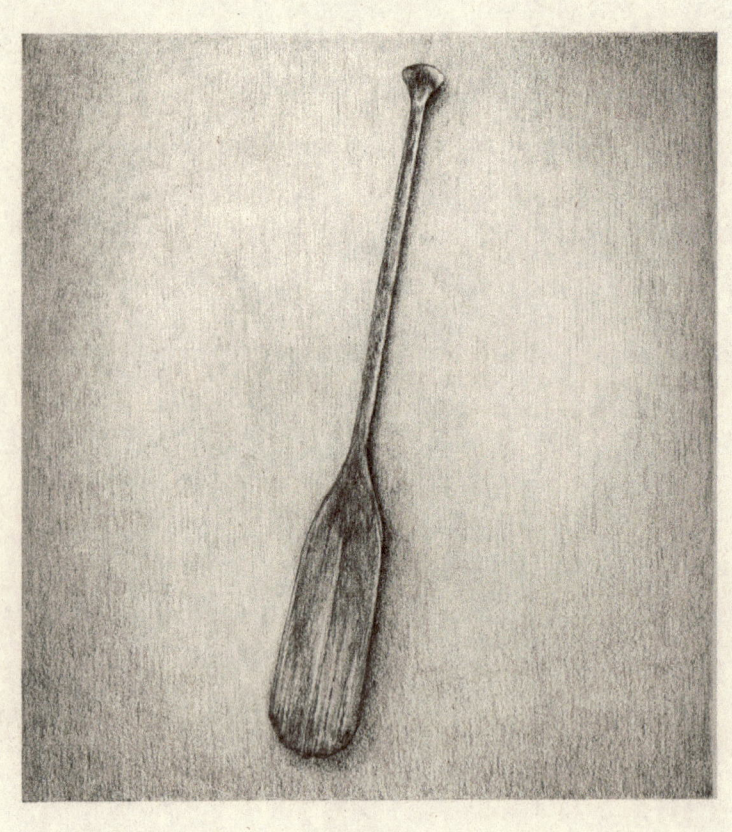

CHAPTER 10

Friendship and Death

We live in a world of wonder and cataclysm.
Everything is borrowed. The worst always happens.

High-end residential building in New York is a small world.
New companies come and go all the time, but only a dozen or
so have been around twenty years. The same is true of the project managers and job supers who work for them. Few stick
around for more than a decade, so the same faces and names pop
up again and again. I can't walk from the East Seventy-seventh
Street subway station to the Metropolitan Museum of Art during business hours without running into someone I've worked
with. It's a world built entirely on reputation and trust. It is also
a perilous world, a world that devours people.

Recently, one of my oldest friends and colleagues, Bo,
stopped by my apartment to go over some drawings for a project
we're working on together. Bo and I first crossed paths when we
were budding woodworkers in Brooklyn. He is two years older
than me. When we met, he was far more skilled. He had trained

at one of the only college-level woodworking schools that existed back then. I had trained in my childhood basement.

Bo is an inch taller than me, significantly slimmer, looser in the
elbows, longer in the face, and just a tad less talkative. He is given
to unprovoked bursts of song, improvising miniature Broad-
way musicals, some with dance numbers, designed to lighten the
mood or make razor-pointed fun at some boss or client who
richly deserves it. We became friends immediately. We have
been friends for almost forty years. We will be friends till death.

Over the years, we've worked at five companies together.
Friends in this business look out for one another. Things go
wrong regularly, and it helps to have someone you care about
nearby.

When we had finished our work that evening, we got to talking about mutual acquaintances. Twenty years ago, the conversation would have been, "I saw So-and-so the other day, they're
working for _____ now. His wife threw him out of the
house!" We know a lot of the same people. We used to say that
there are really only about a hundred guys who build every job
in New York, it's just the logos on the T-shirts that change.

These days, we share different stories. Now, we talk about
who died, and was it cancer, drink, or a heart attack.

"Bo, not a lot of people get out of this business through the
top. Most of them leave in a box."

Bo nodded. "Yeah, it would be nice to move on, on purpose."

We worked on so many jobs together, breathing the same
lead paint, silica, and sawdust. We spent the eighties standing at
table saws with cigarettes hanging off our lips, like it was a point of
pride. We shared bottles of liquor from all over the world, nursed
each other through divorces—two of his and one of mine—and
made each other laugh when our kids wouldn't talk to us. Half

of our friends are dead. But, in all the years we've known each other, we are both happier than we have ever been.

Bo said, "I'm really glad we've been getting together more often."

"Yeah, me, too, Bo."

"It's interesting how some relationships keep working and others are just for a particular time. Like, you meet, and you come together and share things, maybe for a few years, and then that's kind of it. It's really all right. It's not like everything has to last forever."

"Yeah, strange. Sometimes it keeps going and sometimes you turn eleven and Don Knotts just stops being funny."

Bo and I laughed and went out for tequila.

New York City has a reputation for being a cold, lonely place. When I was younger, and more in the habit of feeling sorry for myself, it felt that way to me, too. Now, I think this was more a reflection on me than it was on the city. I liked darker things in those days; sadness and isolation seemed more intense to me, more feeling than hope, compassion, and joy. The world looked unwelcoming; what business did I have pretending it was otherwise?

I had just failed at what every adult in my sphere agreed was the key to success and future happiness. With no high school diploma, I started a freshman college year in New York. Record sharing before the internet was clunky, so my lack of proper credentials went unnoticed. My closest friends were moody and sarcastic. We read depressing books, wore shabby clothes, and concentrated most of our efforts on drinking in tribute to our literary and musical heroes. Nobody, especially young adults, likes to admit that they are in a "phase." Some are so adamant about it that they never let the phase go. But I am not really the brooding poet type. I like things too much, and I find people fascinating. So I quit college, started working for peanuts, and began an education that offers no certificates, only the persistent admonition to learn more.

The first world event I remember from my childhood was the murder of Dr. Martin Luther King, Jr. My father cried while we watched his funeral in the den. I had never seen that before. He told me a great man had been assassinated, a word I'd never heard, and it was a sad day for the world. I was six. As a thirty-year-old man, I read Dr. King's autobiography. Nothing hit me so hard as his writings on agape love, which he defined as "understanding, creative, redemptive goodwill for all men." It's a phrase far better contemplated than explained.

The Selma to Montgomery March, the Children's Crusade, attacking police, cursing townspeople, Bull Connor's dogs, and Dr. King's speeches, I've only seen them after they occurred, but they are the backdrop to my youngest years. I've tried through adulthood to conceive of Dr. King's heroism, the nobility with which he transformed pain. I named my middle son after him. I don't know how a heart can be so great, but I try to hold his definition of agape in my mind when I ponder the question.

A year later, just as summer came, Neil Armstrong and Buzz Aldrin flew to the moon in Apollo 11. Then, they got out and walked on it. It was the first time I was allowed to stay up until midnight. I was seven. We watched in the den. It was the most impossible thing. The TV cut from country to country. Faces from around the world were stretched to the limits of expression in jubilation and astonishment. We ran out to the backyard and looked up at the moon; two men and a strangely shaped ship were actually up there while we gazed from our lawn. My class went to see the rocks they brought back. I built a plastic model of the landing module. As an adult, I took my three sons to the Smithsonian to see the command module the crew traveled in. It is cramped, filled with indecipherable instruments, and covered in char. Two years before Apollo 11's successful mission, three astronauts burned to death in a similar cauldron without ever leaving the launchpad. The courage of these men is unfathomable. I am fifty-nine years old; no one has ever traveled farther; no event has given me such belief in human possibility.

These are the heroes I encountered at the age when I first made my directionless steps into the wider world. Some people have attempted to diminish these men and the things they did—I don't care. The only criticism that has any meaning is criticism among peers. Equal their accomplishments and I will listen. They gave me hope then, and they give me hope now. People are capable of measureless strength, compassion, and courage; they proved it.

Most of us are not heroes. We live our lives on smaller stages, spinning out our days in circles of family, co-workers, lovers, and friends. In New York City, where I still live half my life, we are prodded daily to achieve, to attract, and to succeed, but among one another, if we crave connection, *compassion* and *courage* are our capital. They are all we have to share. A piece of my life can become a piece of your life and a piece of your life can become a piece of mine. The older I get, the less I worry that someone will take me for a fool—I am a fool; I've proven it time and again. And the less I worry what they think, the more people tell me about themselves, because I ask and want to know. When listening, I still have to stop myself regularly from composing the next thing I will say when a silence comes. But when I am able to hold my attention, sometimes my inner listener will ruminate and say along with my companion:

I know how that tastes.

I, too, have tried and failed.

I have been sick.

I have healed.

I have known pain and despair.

I have loved and overcome.

I find the city a much happier place these days. Life can make human beings of us if we let it.

DEBTS

I've always had a soft spot for hippies. I met my first during the ten splendid summers my family spent in that rickety old barn by the Adirondack lake. It was a dusty, spidery paradise, with slanted walls above the beds where we could rub off leaded paint chips with our feet, making fresh peely patterns on which to imagine fiendish faces and birds aflight.

My father had landed a job as the assistant pastor of an island chapel that seemed to draw the entire lake's perimeter to its pews. Each Sunday, the little island was surrounded by canoes, wooden runabouts, and guideboats. The Eagle Island ferry, a school-bus-sized lapstrake lake boat with an inboard diesel and a surrey-style awning, would glide to its designated dock neatly packed with the lake's widows, to a woman carefully wrapped in skirts, gloves, hats, and veils as was the custom of the day. Everyone was Mister This or Miss That; back then, adults didn't have given names where children were concerned.

Everyone, that is, except Elsie.

Elsie Kirkham was the wife of Dr. Kirkham; my parents called him Dunham despite his twenty-five-year seniority, but I would never have considered it. Dr. Kirkham worked at a nearby psychiatric hospital, the kind you used to see in the sixties, where patients aimlessly roamed the grounds in untied bathrobes. I don't know what he did there, and I don't know whether Elsie worked when I knew her. I was a child and never really knew what adults did, at least not for money, because I never saw them do it. The Kirkhams spent their summers all of fifteen miles from their winter home at the hospital, in a darkly stained house of the mid-century Adirondack style, with a big picture window looking out on a glade that led to the lake. The closeness of their dwellings was a ruse; they had been everywhere.

Elsie was the first American-born child of Swedes; she arrived amid World War I on Vinalhaven, one of Maine's rocky island mounds. At the time, the island's chief export was pinkish granite of the highest quality. Elsie herself seemed made of the stuff. While still a teenager, she moved alone to Brooklyn, studied nursing, married Dr. Kirkham, and they were off to a life few have ventured. Their homes were like pages torn from *National Geographic* magazine.

I was too young to know what all the strange and wonderful things were that overflowed from every corner. There was a saddle for a camel, a crooked three-legged stool, drums and clangy bangers of all sorts, rugs and wraps and woven things, all patterned, printed, or festooned with animals, faces, or flowers that anyone could see meant *something*. Unlike in some stuffy natural history museum, you could pick things up in Elsie's living room, turn them over, and look keenly at every detail to decipher how a thing was made, maybe even get an inkling of why and by whom.

Far more important to me than where Elsie had been, or what she had done, was who she was. She had a way of inhabiting herself that I haven't encountered since. Stocky and square, with a raspy smoker's voice, she could convey kindness with a frown, and care with a reprimand. I loved every moment of her.

I don't want to be mistaken; neither the Kirkhams nor my

parents were the types who doted on me in the modern fashion. My childhood was my business to make use of as I wished. Elsie was probably somewhere in the house, finishing the Sunday *Times* crossword in ink, or tending to her explosively colorful gardens (no fish was to be eaten unless its head and entrails were delivered to her for fertilizer). I was scarcely aware of my parents' whereabouts; perhaps they were in the barn recovering from the year's parenting. Once a week, one might be invited into Elsie's back hall to choose an ice cream sandwich or a freeze pop, but except for the occasional family outing, we kids were chased outside, and we were on our own. Fishhooks, matches, knives, axes, air guns, canoes, nails, gasoline, docks, propellers—one or two lectures was provided on the use and dangers of each, then it was up to us to survive or suffer.

Elsie had raised her own children in a similar but even more exotic fashion. David and Marge, her two oldest, began their lives in the South Pacific on a remote island hospital base amid World War II. In my memories, they take on the glow of romantic movie heroes. I never felt that I knew them. Why should I, they were twenty years older than I was, and they wouldn't have been expected to show any more than a passing interest in other people's children. David and Marge lived somewhere else, but from time to time, they would appear. To me, it was more. They made entrances.

They didn't look like the other people on the lake. A scattering of driveway gravel would signal David's arrival. As his motorcycle skidded to a halt, the fringes of his jacket rippled for a few seconds more. Marge and David had guitars, flower garlands, beaded shirts, and hair, lots of uncut, unrestricted hair. These were not adults as I knew them.

It is the luxury of each generation to deride the generations that come just beforehand and afterward, as though faultfinding requires much acuity. In each era's search for purpose, place, and image there will be embarrassing husks accompanying the kernels of insight; intergenerational critics are the laziest and most common kind. I have listened to the hippies' music; I have read from their seminal texts; I've leafed through the photographic record. I know what became of Neal Cassady and Carlos Castaneda; I don't hold it against them.

I was born between generational peaks. I am not a hippie and can never be one, nor do I have any allegiance to the "Me" generation that was the hippies' dissipating comet trail; I have less allegiance still to the "Greed" generation that arose in its wake. But, looking beyond their sometimes-ridiculous costumes and the frailties to which many succumbed, the hippies showed me a few things I wouldn't have considered without them.

David and Marge were smart—fully, conventionally, advanced-degree-bearing smart. Both of them had worked at careers, with offices and salaries and benefits. So, when they chucked it all and turned back to the land, they were sacrificing tangible and, to some, valuable things. After a century of unfettered industrial seizure, most people in this country occupied remarkably rigid rungs on society's social, economic, sexual, ethnic, gender, and religious ladders. Occupants of those rungs have accustomed themselves to looking on those above or below them with suspicion, even fear. David and Marge did their best to abandon the ladder entirely.

Had I never met them, I might never have questioned whether much of life, as people live it, is a prison for all, a psychiatric hospital from which, after donning bathrobes, few manage to escape. Hippies asked some simple questions that planted in me a stubborn skepticism regarding the world I inhabit:

1. Why is a person called "important" when they are wealthy or well known?
2. Do "powerful" people have any useful powers at all?

They set the gears in motion that drove my romantic dreams, not of yachts and fine cigars, but of frozen mountaintops, workshops filled with timeless tools, mornings in the slanted sun of the vegetable garden, and evenings strumming songs by the fire. Those dreams differ little from the life I am still building for myself, but they are joined now by a sense of responsibility for the wider world, and an urge to make good on the promises of life.

Some philosophers and neuroscientists smugly conclude that "free will" is an illusion. For most of the two thousand years since the idea appeared, most people wouldn't have had the means or inclination to give it much thought; illiteracy, bonded labor, and slavery didn't afford a lot of intellectual or social latitude.

Unwavering and unfinished efforts have brought about the world we live in. Presently, about 15 percent of the world's population is illiterate or lives in bondage, at least of the old-fashioned kind. The progress is both palpable and shamefully incomplete.

Individual freedom has been the focus of most struggles; it was the aim of the hippies, and it is still largely the aim of social movements. We are fortunate that, in this country, two of our founding and near sacredly regarded documents are filled with words like "entitle," "equal," "right," "liberty," "protection," and "free," words that bore scant relation to the predicaments of the vast majority of people who lived in the time they were written. I don't know how the men who wrote them could have been unaware of the incongruences their words exposed. I doubt they were.

Those words stand as promises or at least aspirations of how society might someday be organized, opportunity offered, laws enforced, and differences defended. In my opinion, the standards those words set are our country's best hope, its primary virtue, and its most worthy mission. The vision is well stated; the words have been a constant thorn in the side of those who like to declare that we have already arrived, while serving as reminder and inspiration to those whom the promise has not yet reached.

I have been lucky in many ways, but most of all in regard to freedom. In my almost sixty years, no one has ever stood in my way because of what I am. "Free" is not the half of the phrase that has vexed me on my path; my difficulties have come almost entirely from the side of the street marked "Will."

Will is made by what a person does. It is the quality that is inseparable from ability and accomplishment in every sphere. Yet, it is not something that garners much notice. I have always found this peculiar. Without thinking of it as such, people celebrate Will, but they focus on its results, largely ignoring its process. We talk about the extraordinary talent exhibited by a musician or an athlete, but what we are responding to emotionally is the staged performance of the result of Will, the part that can be commodified, the culmination of years of practice and grueling determination to overcome personal doubts, setbacks, and shortcomings, the result of the tedious, triumphant, extraordinary effort we rarely try to see. Will is centered in doing, but its magic is in its ability to transform how and what we are, transformation powerful enough that it is palpable, even inspirational, when we witness Will's results.

About a month after New York City shut down during the pandemic that began in the spring of 2020, I learned from my father that David Kirkham had died of viral pneumonia, just a few miles from where his mother was born more than one hundred years before. From the last time I saw David to the day I learned of his death, the only news I had of him was that he had crashed into a cow on his beloved motorcycle, making a great linguinal mess, not unlike the one I made of my twenties by different means. He gave up riding. Perhaps his sense of duty to his children inspired him to do so.

There was nothing conventional to be done about his death; no one was allowed to travel, and no one held any funerals, so I followed the suggestion of his obituary and had several trees planted in his memory. He was clearly a beloved figure in his community, a teacher, a musician, a mentor. I admire his individual spirit and commend him on the path he chose.

I learned from his obituary that Marge had died before him, but I don't know how. The last time I saw her, she was standing barefoot on a rock in a long blue embroidered dress, wearing a loose crown of freshly picked flowers. She clasped hands with the man she was about to marry. Her story is lost to statistics; his name was Steven Smith. I can still see her inward-cast eyes and the self-effacing smile she wore as the wind whipped her untamed hair.

Marge and David had a younger sister everyone called Tinkey. Even as a child, I was aware that things had gone astray with her. She was possessed of a feral beauty and unfenced demons that adults whispered about when children like me were around. She has three fiercely intelligent daughters who grew up in her madcap house in Marlboro, Vermont. I hope the best has become of them.

By sheer happenstance, David's best friend, Sam Clark, became my first carpentry teacher after my senior year of high school, when I worked with my friend Owen on the rehabilitation of a townhouse in Cambridge's Central Square. Owen went off to fish abalone in Alaska; that's the kind of thing people did in those days. Sam and I were saying our goodbyes when we figured out that we had met at Elsie's house many years before. He still makes cabinets in Plainfield, Vermont, and may even see this someday. I know it's him; there can't be more than one New Hamburger Cabinetworks.

Many of the hippies who showed me the way in my trade have gone. The youngest of them have reached retirement age. Love of craft is difficult to maintain when a body succumbs to failing joints and chronic pain. A select few have gone on to enjoy the thing they set out to do in the first place: They build boats or rockers or instruments, often for free; many teach others the beauty of the processes they love. More than I like to count are lost to the frailties that bedeviled that generation. Alcohol and drug use's euphoric initial promise of Freedom belied the erosion both exact to the Wills of those they ensnare. I witness the grinding awfulness of it still.

I won't judge my elders by their weaknesses, as I hope others won't judge me. I owe the Kirkhams and their kind too great a debt. They inspired me to look beyond the trappings of life, the ornaments and the accumulations, and to put all the energy I can rally into living itself.

OVER THE FALL

By nineteen, I had not developed much sense of responsibility. I know I am not alone in this. There is evidence that young men's brains are not fully developed until about twenty-five. Even if brains worked perfectly at ten, precious few people live unprotected enough lives for them to understand, before twenty, that consequences arise from actions. Parents from my generation were a good deal freer with their reins than those of today. In kindergarten, we were allowed to walk to school with older siblings. A few years later, days of freedom started after breakfast and ended at dinner.

By seven, I would work at my father's table saw unsupervised. By twelve, I had hiked all the high peaks of the Adirondacks. At seventeen, I'd camped in forty-degree-below-zero weather in Wyoming, mountaineered in Alberta, Canada, and traversed the country alone by bus, train, and thumb. Rather than teaching me to value safety, three alpine rescues and a few run-ins with roadside ne'er-do-wells gave me a cocky assurance that I could handle myself in a scrape.

With this as my preparation, I was hired at nineteen to assist in leading monthlong canoe trips in northern Ontario. My older brother and sister had worked for the same outfit. It was run by a gnomelike ex-smokejumper who had learned the ways of adulthood no better than I had. His rules were scant and easily ignored. Training was scanter and concentrated on a destructive fifties camping style that emphasized cutting down as many living trees as possible. The most experienced of his leaders, especially the women, knew it was best, if at all possible, to avoid him entirely.

I arrived in mid-June, expecting to assist one of the older leaders. Two or three days in, Camp Boss told me there weren't enough of them, so he gave me a field promotion, and no raise. A week later, twelve boys and girls, ages fourteen to seventeen, mostly from around New York City, were introduced to me as my "trip." The only advice Camp Boss offered was to tell them I was twenty-one. For the next month, we were to paddle, carry, and live together, covering four or five hundred miles of uninhabited wilderness with no means of emergency communication, seven leaky canoes, and the cheapest food money can buy in bulk. At that age, I couldn't imagine having a better job.

Not one of the kids had ever been in a canoe before. This was an instant advantage, because for the next month, none of them was going anywhere without one. Although I was only two years older than my oldest camper, none of them would have thought for a minute to question why I was in charge. Everything we did was entirely new to them: steering a canoe, pitching a tent, making a fire, cleaning a pot, sharing provisions, sleeping on the ground, bathing in the lake, skinny-dipping; things I had done my whole short life were completely foreign to them. I will confess, although it violates the Benevolent Dictator Code, that watching them take to life in the woods, more enthusiastically every day, was one of the great friendship-forming experiences of my life. Two weeks in, they were running around naked, hoisting canoes onto their shoulders unaided, and walking miles in wet sneakers without complaint. All their opening talk of "What kind of music do you like?" or "Paramus is the best mall" disappeared entirely. Nature worked its magic on them. They were game for almost anything.

I decided to ignore Camp Boss's instructions and take them to one of my favorite out-of-the-way spots. It was only thirty miles or so off our route, but off-limits, if one cares about those things. Our destination was up a narrow winding river, then through a series of four lakes with connecting portages. The lure of the place was a crude cabin, illegally built on Crown land with a small shed containing a wood-fired sauna. You wouldn't guess a bunch of city-raised kids would think much of the idea, but after twenty days in the woods, they thought I was offering Shangri-La.

It wasn't such a demanding trip; the river was gentle, and the portages were short, but on the last portage one of the girls slipped and broke her ankle. It wasn't a bad break, but she had that sugary taste in her mouth that people sometimes get when a bone is injured. I carried her the rest of the way. She was brave and sweet, upset that she was ruining the trip. I put her in the center of my canoe and paddled her to the campsite, now more sick bay than Shangri-La. I smashed a rusty lock off the cabin door and laid her in a bunk. One of the boys had taken a liking to her and brought her dinner and tea. He was even more upset than she was. I wrapped her leg stiffly in a bandage, gave her aspirin and codeine, and we all went to bed. The next morning, I woke up with the sun and fashioned a pair of crutches for her so she could get around. They looked like something out of the *Woodsman's Home Journal;* they pleased her well.

After breakfast, I gathered the group around the map. I showed them that the nearest town was more than fifty miles away by water. The only nearby glimmer of civilization was a lumber road that crossed the river sixteen miles back. The outfit I worked for had an arrangement with a regional floatplane company to evacuate any of Camp Boss's charges should they become incapacitated. I gave everyone tasks for the day, more to boost morale than to accomplish anything, put my assistant in charge, and hopped into a canoe with my strongest paddler, Ari, in the bow. He was two years younger than me, though he thought it was four. He came from Staten Island and bore himself with its underdog swagger. For sixteen miles we paddled hard and carried fast, all the while chatting and learning more about each other than weeks with the group allowed. No one would have taken him for a wilderness boy, with his designer jeans and city accent, but he admitted to loving it and asked how one went about getting the sort of job I had. He had surprised me from the start, and I told him so. I promised to put in a good word for him with Camp Boss and see if anything panned out.

We made it to the lumber road bridge a little after noon. There was no sign of traffic, but the road looked used, so we sat and waited. Three hours later a log truck rolled dustily down the hill toward us. We stood in the middle of the bridge waving passively. It groaned to a halt and two men hopped from its doors. One was enormous—six foot eight or nine—with sloping, expressionless features. The other was small and quick with thick reddish hair capping his craggy, twitching face. "Thanks for stopping," I said. They looked at each other and communicated in guttural bursts. "No English," the small one finally said. We were in the company of a pair of deep-woods French Canadians.

I thought my high school French might be passable, so I offered a few simple sentences explaining why we had stopped them. They seemed to get the gist and responded excitedly in French so unconventional even Sartre wouldn't recognize it. Not a word of it. I reached for the bag I had brought with me and pulled out the section of map I had torn out and on which we were all standing. I handed it to them. I circled the promontory where my group was camping. I produced a scrap of paper with the name of the floatplane company and their telephone number. Each piece of new information made an impression on the pair, so I wrote in hopeful French, "Une jeune fille avec moi s'est cassé la jambe, ICI" with an arrow pointing to the promontory. It didn't have the desired effect, so I made leg-breaking gestures with a hand chop to my leg. The small one drew me closer and pointed hard at the third word in my sentence. He wasn't confused about the leg-breaking bit; he was questioning why there was a girl in the woods. So I crossed out "fille" and wrote "homme" above it. Nods and smiles went around our little circle. They made every indication that they had understood and would see to everything. Then they climbed back aboard their truck and drove over the next hill.

Ari and I looked worriedly at each other. It seemed doubtful that another truck would come, so we slipped back into the canoe and started back upstream, unsure of success. When we arrived at the cabin, our little clan greeted us at the beach. The plane had arrived, the patient had been rescued, and everyone was in good spirits. Ari had become something of a hero. Farther off, a solitary devoted young man couldn't quite bring himself to join in the general cheer; l'amour enfin.

Mutual struggle either separates people or brings them together. This group bonded, probably for life.

As our adventure wound down, I asked if they'd like to take one more side trip to an unusual waterfall I knew not far off our path. By this time, they never said no. So we found our way onto another gentle, winding river, tacked an extra ten miles on to our tally, and made camp by a steeply sloping section of the river. It wasn't really a proper waterfall: The water didn't actually fall; rather, it raced down a narrow chute pitched at a perilous angle forcing the entire contents of the river into a seven-foot-wide torrent. We fell asleep that night while its "Whoosh!" sounded through several octaves.

The next day, I declared our last "rest day." Everyone knew that meant swimming at the base of the "falls," eating any surplus food, and entertaining one another with absurd songs and games. I passed the day quietly at the river's edge. All that water . . . It seemed to me, a pair of paddlers could guide a canoe through it, if only they could avoid a pyramid-shaped rock in the center of the rush, right at the bottom. Several times

I paced the length of the falls, convincing myself of the worthiness of my idea. Finally, near dinnertime, I announced my intention and asked Ari if he thought he'd like to give it a try. The group was encouraging.

So Ari and I carried a canoe to the top of the chute and circled around a few times. In truth, I had limited experience running rapids. What I had was years of experience in being foolhardy. My instructions were simple: The second the current took us, we would both draw left with all our might, thus avoiding the rock and proving ourselves incredible. Ari listened closely and nodded that he understood. I pointed the canoe downstream and stood up in the back to survey our path. It was a pointless posture; the chute fell off too steeply to see. Ari reached the crest of the falls and the canoe lurched forward in the current, pulling me down into the stern seat. Together we flailed madly on the left side of the canoe trying to move it in that direction. Our paddles had no effect. The canoe careened down the center of the chute, straight at the pointed rock below. The hard keel of the bow crashed directly into it. Ari was launched out of his seat and flew headfirst over the rock into the

pooling water below. With his ballast gone, the canoe levered broadside and capsized. My legs smashed against the rock as I flew past and I was plunged into the swirling pool next to him. I looked back to see the canoe wrapping itself like tinfoil around the pyramid rock, creakily folding into a slender C. I swam toward Ari and followed him onto the shore. My legs were banged and bruised, but nothing was broken. He pulled himself into a squat, unscathed.

We walked back upstream to have a look at the canoe. We both let ourselves into the current and gave it our best tugs, but with tons of water pressing against it, it was never going to budge. I'm sure it's there still.

I've seen expressions like those on the faces of my campers several times in my life. There is a combination of awe, terror, and profound relief that few events can bring about. I never asked Ari how he felt about what we did. At the very least, he was happy to be alive.

In two more days, we returned to base camp, minus one canoe and one rescued camper. I was concerned it might mean the end of my employment, but Camp Boss greeted us jovially, showing us where he had hung the crutches on a cabin wall, a fine example of vintage woodcraft. My lovelorn camper was reunited with his paramour. And on my recommendation, Ari was promised a place as an assistant for the following year, his wild man status now firmly established.

Many years later, while looking up old friends, I came across a news story. At a railroad crossing on Long Island, the driver of a Jeep foolishly drove around lowered signal gates and was struck by an oncoming train. Ari was pronounced dead at the scene. A police officer found his six-month-old daughter strapped in the car seat behind him. There wasn't a scratch on her.

CROOKED TONY

Before Tony died, he asked me to take his dog, Minnie. She was a special breed I've only seen in New York. I call it Brooklyn Junkyard: part German shepherd, part Lab, and a big helping of who knows what. I don't understand how breeding that is so random can produce so many dogs that look alike. Twenty years after taking her, I see unkempt mutts that remind me of her from time to time when I walk through the city's manufacturing districts, though less so than I used to. Pit bulls are everyone's favorite castoff these days; the unwanted denizens of New York's vacant lots now have squarer jaws and shorter legs beneath their barrel chests. Minnie was smart. I never trained her a bit, but I could say, "Get off the couch" or "Go get your ball," and she'd do it. Thanks to Tony, she was sweet, always tolerant of the less-than-gentle attentions my three sons lavished on her.

Tony and I had first worked together on the Snail job and later at Robin Williams's place. He was wiry and strong from years spent biking everywhere, not in the annoying cyclist way; he just went everywhere by bicycle, in blue jeans. He usually sported a slightly startled grin, and his hair would change shape and color from time to time. When we first met twenty-eight years ago, he was my foreman. He taught me the importance of doing rough work as neatly and thoughtfully as the fancy stuff. To him it showed care and competence to get every step of the work right.

After decades of physical work, Tony started to have regular pain in his back and saw a series of doctors who told him it was trouble with his discs. They were completely off the mark, but he had no way of knowing that. He underwent a spinal fusion and spent months in a body cast. He lived nearby, so I'd visit and bring him things he needed. Even in his cast, he was good-humored, ready with a laugh. But he was terrible at being sedentary. He began making pottery, which he had studied in college. He made pots and vases of all sorts; gray and muddy, they were scattered about his apartment, uninterested in finding formal homes.

A few months after surgery, he hadn't properly recovered. His spine was fused, but the pain had gotten worse. New doctors were consulted; tests were done, and he was finally diagnosed with multiple myeloma, which in those days was nearly untreatable. He could still get around, just not easily, so I invited him to come to the Adirondacks with my family to stay for two weeks in our little lakeside cabin. Tony was thrilled. He was a city kid, and everything woodsy was an adventure to him—the boat, the gaslights, the water pump—he liked it all. Tony swam, fished, and swung around in the hammock in utter contentment. Then one of the kids started throwing up and got a terrible case of diarrhea. Then we all got it. Tony got the worst of it. I felt terrible for him; he was already in constant pain and now my family had given him this. But he was indomitable in his insistence on enjoying himself. In a few days he was able to keep food down.

One morning, my family was taking the boat to town for supplies. Tony asked if he could be dropped off at the beach on the opposite shore so he could climb the mountain he'd been gazing at across the lake. My wife packed him a lunch and we deposited him on some rocks at the sand's edge. Ampersand Mountain's summit is an eight-mile round trip from the beach, half of it muddy and steep. I've climbed it fifty times in my life but never with a fused spine and cancer. I didn't know if Tony would make it, but he wanted to try, and I've never felt it was my place to urge safety when little was left to be saved.

When we returned from town and pulled the boat up to the beach, there he was. He was muddy, limping, and grinning from ear to ear. It was the first time I've seen someone actually beam with pride.

Two months later, Tony started making preparations for death. He was flat broke. Sloan Kettering had taken him on for compassionate care, out of kindness and because his condition was rare and thereby intriguing. But he hadn't worked in more than a year; he had a daughter and lived in Brooklyn. Money had run out; even people who know they're going to die have to eat. Tony lived in an old apartment that he'd fixed up, with a deck overlooking the Gowanus Canal, a Superfund site that had tested positive for cholera. The canal and all the lead paint and construction crap he'd inhaled through the years are probably what did him in. The time had come for him to move. He threw a party. Scores of people came. We brought food and most everyone bought one of the ceramic pots he'd made. Mine was from a series of vases that were twisted and crooked, as Tony had become.

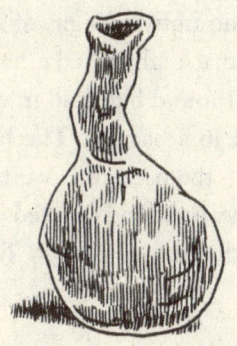

The next day, a few of the guys from work joined me in moving his stuff out of the apartment. It was brutally hot. We were all covered in sweat, struggling with a couch on the stairs, when one of our company's project managers showed up. "Oh great, we could really use an extra hand," I said with relief. He looked at me blankly. "Oh sorry, I'm here for some quality time with Tony," he said as he climbed past us. I have never even entertained forgiving him for that.

Tony went into the hospital and fell apart quickly. I visited him twice. The first time his mother was there. She was a compact, aging Italian woman with the remnants of an accent. We had never met before. She thanked me over and over for looking out for him and hugged me tightly after Tony and I had talked our way through visiting hours. The next time I went, Tony was comatose. I talked to him and held his hand. I told him I loved him and that the kids loved Minnie. He died the next day.

His mother held a Catholic funeral for him in outer Brooklyn and buried him nearby. The priest had never met Tony. It had been a long time since any priest had. The service was incongruous with the man I knew, but I could see that it gave his mother solace knowing that so many people loved her son.

Minnie lived with us for seven years after that. One morning she was out of sorts and wouldn't eat. We watched her through the day. There were no outward signs of illness; she just wasn't herself. We decided that night to take her to the vet the next day. In the morning, I found her dead in our first-floor powder room. I wrapped her in a blanket. The boys and I took turns digging a deep hole in the backyard. We laid her in her blanket in the hole, said a few words, and covered her with dirt. She was a good dog. I cried that day as though Tony had died all over again.

———————

The first time Pauline pulled me up short, we were chatting about some apartment I was working on. I made a comment about the owners' "Oriental" art collection.

"We don't say that anymore," she told me, smiling.

"Don't say what?"

"Oriental—we say Asian."

"Really, how come?"

"It's offensive; Asians like to be called Asians."

I'd had no idea. I've never kept abreast of much of anything. She was kind but firm about it, so I took it to heart and never used the term again.

Pauline worked as a secretary at a general contracting outfit I worked with for five years. She was smart, pretty, and one of the only people in the office who didn't care what my position was. She liked me and treated me in a way that let me know it. When she didn't like someone, they were made equally aware of that circumstance. I didn't get to the office much, but whenever I did, we'd slink out back for a cigarette and chat about things. I was married but enjoyed having a crush on her that would go nowhere. She may have felt the same way—I never asked her. We were friends.

On one of our breaks, I told her, "You ought to be a project manager; you'd be better than half the ones they have here." She thought I was just blowing smoke; she pointed out that she didn't know anything about construction. "No, I really mean it. You're smart. You're organized. You're not scared of anybody. You know how to talk to people. That's way more important than what you know." She told me she'd think about it. She did. She started taking courses at night after work. Eventually, with repeated pestering, the boss promoted her to assistant project manager. I couldn't have been more proud.

Pauline grew up in Queens; her parents both came from China. Her public school teenage years, rife with mistreatment, toughened her; she had a back-sassing armor that protected and complemented her devotion to those for whom she cared. Four of us who worked for the company, two men and two women, became close friends. We formed a birthday club, taking one another out each year, exchanging gifts, and keeping up with one another's lives. I don't remember how it started. Likely Pauline or Anette noticed that Cliff and I had birthdays in the same month, and suggested we celebrate. No one would have guessed when we started that the tradition would carry on for almost twenty years. Together, we celebrated two marriages and three divorces, none between any of us, thankfully.

I left the company we all worked for after five years. Cliff and Anette followed me to my new job, but Pauline stayed put. She had a wider loyal streak than the rest of us. It took me ten years to talk her into working with me again. By the time I did, I had moved on two more times. I was mid-divorce and looking for shelter. Pauline had just suffered a miscarriage, a daughter, Grace, and was diagnosed with breast cancer shortly thereafter.

My job needed a project manager a few days a week, and there was no one else I knew who could turn a drama that featured Park Avenue, summer rules, divorce, and cancer into a cozy bonding experience. Neither of us had the energy to care much that our work was unappreciated. Pauline underwent one treatment after another; remissions and relapses followed one another with regularity. The anguish of my divorce, the havoc it wreaked on my relationship with my sons, and the emotional calisthenics of my love life formed the tempests of my weeks. When that job was finished, we were joined by our friend Cliff on the largest project any of us had done. He had been diagnosed not much earlier with multiple myeloma, the same thing that had killed Tony. I warrant that three people were never less emotionally suited to carry out one of Manhattan's most ambitious renovations. Circumstances weren't improved when I had a bicycle accident, breaking my hip on the way to work. I tried to ignore it, and rode the few remaining blocks to the jobsite, but Pauline could see that I was hurt worse than Advil could handle. She found me a broom for a crutch and took me in a cab to the emergency room. The next day, after a surgeon had bolted my bones back together, she and her husband drove me to my sister's house in the Catskills to recuperate. A few days later I was back at work. We were a ragged band, but we had known one another for a long time; work was our battlefield and we were foxhole friends. The three of us, and Jesus, whose mother was dying, worked like we always did, in the only way we knew how. Work was our comfort, and our companionship was our strength.

Most days, we would leave the site for lunch and head to a nearby restaurant that Pauline nicknamed "Shitty Chinese." It was a standard New York, First Avenue place; in truth, the food was fine. We had the same waiter, Sam, every day. He would bring Pauline special dishes she ordered off-menu, and he would make me a custom soup that was too spicy for everyone else. We could ditch the design team and talk freely there. We told jokes that anyone would have categorized as "too soon." But we were all in peril, and all in pain, and we loved one another in an easy way that made all of our burdens lighter.

A few months after the job finished, Pauline had a relapse and started seeking out alternative treatments. We saw each other for lunch once a month or so.

An intense romantic relationship in which I had placed the

highest hopes fell apart one early afternoon; I drove to Shitty Chinese for lunch alone trying to feel better, but I couldn't eat; I could barely breathe. I called Pauline so I could feel that someone cared. She told me how dumb I had been; she never thought it would have worked out with that girl. We had a laugh about it, and I made it through the afternoon.

A year and a half passed. We kept up our lunches in a less regular way. My heart mended slowly, and Pauline looked healthier and stronger. We let a few months go by without seeing each other. One morning, I got on the elevator in my building, pushed the Lobby button, and said out loud, "Looby." Pauline and I had always laughed because someone had spray-painted "Looby" outside the elevators at our last job. It was fifth-grade funny, but it tickled us every time. I texted her and asked when we could have lunch. She told me that six lesions had been found in her brain and she had surgery scheduled in two days to relieve the pressure.

I didn't hear from her for a few days; I would send little prodding texts telling her I missed her and hoped she was feeling better. I finally got a text from her husband, Victor, her childhood love. She had taken a turn for the worse. Her latest scans found cancer in her liver, peritoneum, and around several vertebrae. He told me she was awake but weak and unable to eat. I asked if I could visit, and he told me she would like that.

Anette and I talked on the phone. I arranged to pick her up and drive her to Connecticut. Jesus agreed to come the same day. There were so many people, friends and family, who wanted to see Pauline; Victor made a schedule and arranged visits for everyone she wanted to see, politely keeping away the rest.

When we got to their house, Victor hugged us at the door. Soft Buddhist chants played in the living room next to a hospital bed where Pauline lay. She was so thin; her color had left her;

there was a six-inch semicircular scar on the side of her head; she was unconscious, almost unrecognizable. We held her hands, touched her legs and feet, and talked with her, Victor, and her sister for two hours, until it was time for the next group to arrive. Victor was angry, and so terribly sad. He was trying to get Pauline's medical interventions removed so she could die without prolonged suffering, but her family still held out hope. It made it even harder for him. I told him I thought he was right; Pauline was brave and loved life, but she loved the living kind, not the kind that held on needlessly when hope was truly gone. We gathered our coats and shoes, exchanged more hugs, and all turned to Pauline to say, "So long." Until then, she had only moved in uncontrolled spasms, but as we left, she raised her left hand weakly and waved goodbye to us without opening her eyes.

Anette cried in the car on the way home. She wanted to arrange another visit for later in the week. I didn't think there was going to be another visit.

The next day Victor sent word that Pauline had died that afternoon. It had been such a long and rough road for her, yet I can only think of her smiling, asking after her friends, making jokes, and chastising me for all the flaws she forgave in me.

EXTINCTION

I'm a dinosaur.

Of all the rooms I've visited, only a few have emblazoned themselves on my memory: the knave of Chartres Cathedral, the entry hall to the Smithsonian's National Air and Space Museum, the Main Concourse at Grand Central Terminal, the Metropolitan Opera House, Ryman Auditorium. Some of these stick with me for their grandeur, some for what they house. The one room that stays with me for both is the Dinosaur Hall at the Carnegie Museum of Natural History in Pittsburgh.

When I was too young to find my own way there, my mother would take me. As we strolled, I learned how to negotiate traffic signals at the largest intersection on our route; I heard for the first time about sightlessness as we passed the school for the blind; and I had my face smudged clean after ice cream with the corner of my mother's handkerchief and a dab of her saliva. After a half dozen of these excursions, I was deemed a responsible navigator. So my seven-year-old self would set off with a notebook tucked under my arm and a pencil in my shirt pocket to sketch the continent's unearthed giants.

Mr. Carnegie had seen to it that I could get in for free, and Joe, a friendly security guard, would keep an eye on me that I might not wind up in a stranger's trunk. The marbled Dinosaur Hall was majestically made, tall and loomy, with oversized reassembled beasts in every direction. The stone floors rang and echoed with my steps; I pronounced their magnificent names as I passed each display: *Brontosaurus, Triceratops, Stegosaurus, Tyrannosaurus rex.* Occasionally, I would stop and try a few pencil strokes, quickly frustrated by my subjects' bony complexity.

There weren't just dinosaurs. Before humans wandered over from Asia, North America's forests hid great lumbering bears, fierce snaggletoothed cats, and man-sized fish. They were posed around the room, some taxidermied and furry, others skeletal and menacing. There were sloths and beavers that could feed a family for a year, and once people arrived, likely did. In the middle of the hall, a lone nearly naked huntsman faced off with a lion, looking unlikely to prevail with his atlatl and spear. With only wits, nimble fingers, and a tongue made for tattling, it's a puzzle he was the sole survivor in the room. The stuffed mammals stood in memoriam, the evolutionary victims of all creatures' nagging need for dinner. Looking about, even a seven-year-old could deduce that, sooner or later, we would be joining them.

I toured the museum time and again. Each visit began in the Dinosaur Hall; few young men of my generation have any inborn resistance to their charms, but once sated, an exploratory disposition took me all around the sprawling place. The top floor was home to anthropological dioramas that showed ways of living far more diverse than those I knew, starting with a band of Neanderthals, supposed as our grandparents at the time, and leading, like the pages of *National Geographic,* through African tribes, nomadic herdsmen, headhunters, sun-dancing Lakota, and finally, the first Europeans to reach these shores. The scenes had the air of a time just passed, as though I had arrived late and missed everything. Perhaps my grandfather, born just within the twentieth century, could have witnessed such things in person; perhaps he knew someone who had trekked or sailed too far and had lost his head to discovery. Our times were no tamer: Young men were being sent off to fight enemies that seemed exotic and unknowable; Pittsburgh's streets broke out in riots that were quelled by force. But, to my young mind, struggle seemed fairer and more noble for the immobile diorama dwellers who fashioned implements made of sticks, flint, and bone.

Heavy with questions, I began my descent from floor to floor. I would breeze by the plants and protozoans, stop briefly to wonder why so many worms, and throw sidelong glances at the dusty butterflies and birds—the mammals were my object. Frozen in their forests, deserts, riverbanks, and plains, they felt more alive than their pitiable cousins imprisoned in Pittsburgh's concrete zoo.

Mr. Carnegie achieved in me his expressed aim. In his day, these bones, carcasses, and sarcophagi were sought, stolen, and assembled to show the vastness, variety, and genetic order of the nearly conquered world, pushing aside religion's fables with the clear evidence of Nature's primacy. The entire museum, laid out like a map of cultural, genealogical, biological, and evolutionary supremacy, was an homage to this order. Huge gaps remained clothed in question marks, but that was Science's beauty: There was always another discovery to be made, ever new worlds to explore. His was an age of surety in the faith that people had found their natural place as custodians of the planet. Every image of Mr. Carnegie shows him standing straight and proud, his large, inquisitive brain cradled aloft in starched collar and bow. He had every reason to be self-satisfied. He had traipsed society's entire length in one life span, from bobbin boy to robber baron. He was, for a time, the richest man in the world and its greatest philanthropist, pledging to unlash from caste-ordained moorings little boys just like me. The museum's edifice may have looked like a classical temple, but in its own time, it was a monument to the revolutionary usurpation of a world once ruled by the right of heredity. The bronze weapons of this new order loom at its entrance: Bach for music, Michelangelo for art, Shakespeare for literature, and Galileo for science. Carved into the building's frieze are the names of those who fleshed out the pantheon, like phyla of the four kingdoms. Certainly, hosts of undiscovered, unrecognized, and shunned names were omitted, but evolution's march would someday allow their inclusion. Knowledge and understanding were sure to undo what tribalism, superstition, and small-mindedness had ordained. Mr. Carnegie was the long-dead living proof.

Every few years, the art side of the museum would put on an ambitious show of modern sculpture and painting called the Carnegie International. The first such exhibition was mounted during Mr. Carnegie's lifetime, and it has carried on through mine. It was meant to be progressive, but no one in his time could possibly have imagined what it would become by mine. Among the galleries of yellowed works, all tethered to Brunelleschi's vanishing point and naturalism's pre-photographic hope of accurate depiction, the riot of block-colored, disproportionate art was half insult, half war cry. It meant to throw aside the rigidity of rationalism, positing meaning in primordial gesture, form, and color. Rationalism's formality was undone by the intervening century of war and horror. It had thrown art back into the elemental and brutal world of the prehistoric, back to an unspoiled romance with life before the stain of civilization.

How many waves of the "new" do people see in a lifetime? The Mesozoic era lasted nearly two hundred million years. The Big Band era lasted just long enough for my great-uncle to learn to dance. Are evolution's machinations set to refine and improve, or are they thoughtless grinding gears of destruction that allow only a select few to thrive?

When the brain outstripped the stomach as the governing organ of our bodies, twin paths of mutual understanding and mutual annihilation stretched into the future for our species, with seeds firmly planted in us for both. Vanishing points are entirely imaginary and starkly real. With every step taken down our paths, the conjoined parallels remain visible on the horizon, yet recede in the exact measure of the distance we travel. No matter how we ache for great conjunctions—joy to sorrow, life to death, reason to mystery, beauty to horror—the parallels pry themselves apart; we remain incomplete, and then we are gone.

CHAPTER 11

Architecture and Art

We are what we have done.

I've never wanted to be an architect. I have always loved to build things, and my experience in the industry has shown me that architects don't get to build. The architects I know divide their time among regulatory filings, marketing, customer relations, small business administration, contractor oversight, and drawing. Building isn't on that list, so architecture is something I never wanted to do. It amuses me that the word itself, "architect," means "master builder," as that is something most of them never do, at least not in my end of the business. Decades of university theorizing have systematically divorced the profession from the physical act of construction so that a more accurate modern definition of architect might be "person possessing an interest in the compelling nature of built structures who, having earned a college degree and an elusive license, believe they have the authority to tell builders what to do."

Several architects I've worked with have confided in me that not even they are sure that what they do can be called architecture. High-end residential architecture has no need of concepts like shelter, sustainability, value, or material appropriateness, unless they can be exploited as fashionable. It took me fifteen years of work to understand this, distracted as I was by dreams of craftsmanship, something *I* find too few chances to pursue.

For those who can afford it at this level, architecture's primary purpose is to flatter the self-image. Of course, there is nothing new about this, but it sounds shocking and hollow to say it directly. It explains why so many beautiful things in this world are hoarded by people who don't seem to have much appreciation for them.

A CHAPEL FOR ONE

I have worked on at least a dozen projects that are completely forgotten. There was one, maybe twenty-five years ago, from which all that I can remember is the name of the designer and a single photograph in a magazine. Everything else has disappeared: whose house it was, what we built, which workers were there, what year it was—all gone. But it was on this project that I finally understood high-end architecture's purpose.

The project designer's name has stuck with me through the years; it was one of those three-part, English-sounding names that makes you wonder whether it might have been invented to cover a less-than-well-born beginning. I don't remember another thing about him, just that name.

One day, around noon, we had all stopped for lunch, and one of the guys was flipping through a décor magazine featuring a southern country estate designed by our English-sounding fellow. The estate owner was one of the better-off billionaires at the time, and his home away from home had the gardens and grandeur to prove it. None of this was of interest to my co-worker. He was looking for a photo he'd found of the owner's wife. He worked his way through a few rounds of "continued on page . . ." until he could produce it and hold it up for the crew to admire. "You know who that is?" he asked. Our blank stares betrayed that we did not. "That's _____ _____!" Yet another detail I have forgotten. No flashes of enlightenment appeared on any of the gathered faces. "The porno actress! C'mon!" This was long before PCs and smartphones, so the assembled crowd was not as well versed in the oeuvre as they might be today. "I swear, I read that the dude saw her in a movie, got a hold of her agent, and had her flown down to Palm Beach. Like, six months later, they were married. Fuck, rich dudes, right?"

I have never had any idea if any of this was true or not and researching it now would spoil it for me. It was a compelling photograph. A lovely young woman stood in the foreground, neatly corseted in a mid-nineteenth-century gown, her hair done in ringlets, her cheeks touched with rouge. Behind her stood a regal assemblage of service people elevating her all the more in their semicircular embrace. All were arrayed before the stretching limbs of a grand old oak whose leaves shaded them while obscuring the façade of a miniature stone chapel. The scene was more idyllic than natural laws can produce.

The accompanying text provided all the further illumination the photo needed. The mistress of the house loved the history of the region and had hoped to recapture some of its romance. She allowed that the servants enjoyed dressing up in period livery for formal affairs. The gardens had all been planted at her direction, and the little chapel's stones had been carefully stacked so that she might have a private spiritual retreat from the demands of her daily burden. A young local padre had been persuaded to come by each Sunday to offer her Communion. Picturing her solemnly kneeling before him in the chapel's shadows was a snap.

Patrons of the Renaissance's greatest painters were outmatched by her at every turn. Their images are shoehorned into the margins of devotional masterworks, but with all their powers, they never orchestrated a more perfect marriage of piety and porn.

I would make a poor arbiter of taste. I don't seem to have the knack for it. Much of what I've been asked to build in my life was considered stylish as I built it; some has held up over time, most hasn't. Trends never serve anyone well; they're mainly marketing schemes, efforts to make a splash. I gave up on them at an early age after some failed attempts at fitting in.

As though adolescence weren't sufficiently awkward, the fashions of my puberty made it acceptable to wear a russet blazer over a lizard-tongue-collared polyester shirt bearing scenes of wild western horses. All of this topped a pair of flare-legged burgundy corduroy trousers that drew the eye downward to my suede crepe-soled platform dancing shoes. I was decked out for a Bar Mitzvah party, but I should have been dragged to juvenile court.

To this day, in my own home, I assemble rooms of furniture and décor in ways others find haphazard, even vexing. Lately, I do what pleases me without apology and I encourage others to do the same, critics be damned. Everyone has to live in their home; no one has to live with critics.

For those who think they might want to brave the critics and design tomorrow's world, I offer the following discourse:

A BRIEF HISTORY OF ARCHITECTURE

Five hundred years ago, architects like Andrea Palladio were actual builders. At thirteen, Palladio apprenticed as a stonemason and bricklayer. He practiced that trade for twenty years, until he went to Rome to see its classical ruins. The trip forever altered his vision. Upon his return, he began building some of the most serene houses, palaces, and churches in all of Italy. He became an architect, a supreme one. My eyes find it difficult to stop looking at the things he built. His *Four Books of Architecture* often read like pleasant conversations with an older tradesman who is happy to share sound practical advice.

A few hundred years later a Swiss fellow came along whom we call "the Father of Modern Architecture." Le Corbusier (not his real name) trained briefly in watch manufacturing, moved on to drawing and painting, and eventually took an interest in architecture and urban planning. Like many that followed, one of his first major commissions was for his parents. The house went so far over budget that the unfortunate pair had to abandon it. "Le Corbu" hated decoration, natural color, handmade objects, tradition, all things non-European, and children. If a building's quick and constant decay is any indication of its builder's skill, Le Corbu had little to none. He wrote like a bad French poet, conducted himself like an imperious buffoon, and has been one of architecture's great luminaries and role models ever since. I find it difficult to look at his buildings, but if you'd like to, I can recommend Villa Savoye, the prototype for every soul-crushing suburban office park that followed, and the Convent of Sainte Marie de La Tourette, a "religious" building so satanically ugly the lichen on its walls look embarrassed by their slow progress at chewing it to pieces.

Palladio studied and took advantage of technologies that were largely unchanged since Roman times. Le Corbusier's life fell squarely on the cusp of humanity's most transformative era.

Industrialism undid thousands of years of agrarian human history through means that were most welcome, like penicillin, anesthesia, and vaccines, and less benevolent, like mechanized war, ecological degradation, and the destruction of local traditional cultures.

Builders suddenly had a host of new materials at their disposal, among them cold-rolled steel, reinforced concrete, and plate glass. They no longer had to stack stones and join wood to make a structure stand. Steamships, some of them bigger than the world's tallest buildings, chugged across the oceans; automobiles replaced horses; men rode ragtag mechanized kites into the sky. Freed from its allegiance to nobility and religion, architecture could stretch its wings and celebrate the power of the Mechanized Age. Those early modern architects must have felt like superheroes. Some even wore capes. Their meeker modern-day descendants satisfy themselves with Le Corbu's spectacles.

While reworking the skyline and most of daily life, industrialism re-sorted the social classes. As factories replaced workshops, management separated itself from labor, robbing it of its independence. The result is a world where two-thirds of us come home dirty every night, and one-third comes home clean. One need only ride New York's subways at rush hour or survey the contents of the city's cabs to see who is who.

With the help of the early twentieth century's burgeoning university system, architects left the dirty working world behind and joined the clean one. It was architecture's loss and it continues apace. Drafters' hands are no longer smudged with ink and graphite. Sketching by hand, one of my most useful skills, has been cast off as anachronistic. Eager young architects excitedly tell me about their offices' shiny new 3D printers, used for "rapid prototyping," and all I can think of is the few remaining carvers I know who can do the work in a fifth of the time, and far more artfully. In twenty years, I expect most of the young architects I know will be much like their employers, owners who manage to keep a small business's doors open without ever really understanding how those doors work.

Who didn't dream as a young person of a life of accomplishment and accolades? I certainly wasn't spared. I can't blame the young crowd for seeing the symphonic romance of building beautiful things, and secretly pining to be Bernstein rather than second oboe. But they should know that oboe players know a few things Bernstein never understood.

Like every earnest endeavor, building is a daunting responsibility. We are the only species that so brutally scars the landscape. The best examples of what we've made share a few common characteristics:

1. They are safe, at least for their time and place. If the natural elements have swept them away or too many people have tumbled over the railing, something went wrong.
2. People don't despise them. Communities are powerful things. When a structure offends too many of its members, it is unlikely to last.
3. Craft is evident throughout.
4. They know their place. The best buildings feel at home where they sit, whether a monastery clinging to a cliffside or a thatched roof cottage nestled in a glade. The most

glorious cathedrals express power and awe as effectively as the quaintest fishing huts express determination and stubborn humility.

In our time, one more characteristic must be added:

5. Good buildings don't ruin the Earth.

In my four decades of work, I have never built anything that fits those criteria. For the first time in my career, I am working on a project that tries.

We return at last to the 180-year-old side-by-side townhouses in leafy brownstone Brooklyn. Atop the just-completed concrete foundation, their crooked timber frames sit, secure and safe, braced from within by an internal steel skeleton far stronger than the creaky wooden bones it supports. The houses are the same old tinderboxes they have always been, but the fireplaces and gas stoves are gone, so there is less to set them alight. Progress has been made. If we had demolished the houses and rebuilt them from the ground up, stricter codes would have been triggered, and the modern definition of "safe" would have been met, but that would have put us squarely in opposition to community opinion.

New Yorkers love old townhouses like Belgians love cobblestones. As unaffordability spreads deeper into each borough, new generations of owners tempt themselves into believing that their not-exactly-majestic homes might be historic. A block away from our side-by-sides is an avenue of detached mansions standing testament to the ostentation Victorian New Yorkers valued. Europe's finest displaced craftsmen built those homes. They were as eager to display their hard-won skills as the original owners were to showcase them.

Running behind that avenue is the street where we are working. Low roofs shelter stables and servants' quarters, dotted by an occasional merchant's or butler's house mimicking in miniature the fancier fare found a block away. These homes were built by different crews, with cruder methods and cheaper materials. They are historic more for their social context than for their architectural merits. But in a neighborhood where people pay millions of dollars to live in a stable, God help the new homeowner who tries to tear a townhouse down. Community spirit is just as likely to defend the perception of provenance as it is to condemn an obvious eyesore.

The twin townhouses have been saved. Now it is time to restore their façades and to fit out their interiors in the spare, speckled style of our time. Woodworking skills have been preserved well enough that we can provide period exteriors that surpass the quality of the originals. Modern paints, glues, membranes, and sealants will ensure that, with proper upkeep, the houses will last well beyond my grandchildren's life span. The interior's future is not so assured. There are moments of flair; once again, sculptural staircases take center stage. But in fifty years, most of the rest of the house is likely to fall on future eyes the same way the sunken living rooms and paneled dens of the seventies fall on ours. The design world's currency has rarely been timelessness.

I can't rightly call most of what we do craft. Modern design and budgets leave little room for it. We have sunk so much money into shoring up the few original sticks that will remain in these houses, and so many months correcting their geometric eccentricities, that craftsmanship's tempo and expense would tax any client's patience beyond tolerance. Aside from a mosaic bathroom and sweeping staircases, from here out, we will build these homes expeditiously, relying on unconventional wallpapers and funky custom countertops to provide whatever visual punch we deliver. The architects are pulling one time-honored ace from their sleeves for visual grandeur: double-height kitchen/ living room areas, lined on both sides with windows that climb to the ceiling. That should give the photographers something to play with.

The architects refer to their design concept as a "palimpsest," a word nearly everyone has to look up when they come across it, which is half the point. I've had little call to scrape and reuse old parchments myself, so we may all be forgiven our lack of culture. I'm not sure the metaphor is apt. It seems that what we're doing is more like scraping the vestiges of soft tissue from old dinosaur bones, stretching enough modern-day lizard skin over the outside to make a convincing paleo-facsimile, then installing new innards made of enough modern heating, cooling, and lighting systems to bring it all back to life. At long last, my dictionary fails me.

The houses will still look right at home in the neighborhood. We'd be fined if they didn't.

We've reached our final architectural question: Will these buildings ruin the Earth?

For any permanent structure, the answer will be a matter of degree. In this case, the addition of a new foundation made from one hundred cubic yards of concrete, and the thirty-three-ton steel frame that it supports, guarantees that our little project has already added 150,000 pounds of CO_2 to the atmosphere. When we ruin the Earth, we builders go big. New technologies like mass timber and carbon-neutral concrete have recently become available; that impact could have been eliminated. Once the three thousand or so building departments in this country meet and approve the use of these products, the problem will be solved forthwith. I understand that these materials seem expensive now. Solar power was once prohibitive, too, and now it's the cheapest electricity available, ever.

If I enumerated all the ways we are using resources, burning fuel, and creating waste to build two modest single-family homes in Brooklyn, a pall would drop over this chapter that I might never find the levity to lift. I will stick with the good news.

I'm unable to think of a single detrimental action we are taking in Brooklyn for which a much less harmful or even beneficial alternative does not already exist. Only two things stand in their way: Purveyors of harmful technologies ferociously defend the profits they enjoy from the punishments they sell, and precious few educations are rational enough that their possessors might recognize and adopt good solutions when they come along. A hopeful attitude toward the future doesn't exclude a critical view of the way things are.

The Brooklyn twins are trying. Both will earn "Passive House" certification. Various groups have tried and failed to trademark the phrase in this country. Much like "green" or "organic," the term is a marketing tool for salespeople. All three labels conjure up images of too-serene families clad in billowing linen practicing yoga in the solarium. Most of my co-workers wouldn't rush to buy Ford's new "Passive Pickup" or feel flattered if their paramour introduced them as "my Passive Boyfriend." It doesn't hurt to introduce a beneficial idea with a dash of élan. Even something as tired as "high performance" would be an improvement.

The principles we're following are simple enough:

- Insulate and seal the exterior of the house as thoughtfully and aggressively as possible.
- Make sure all doors and windows are as miserly with heat transfer as the owner can afford.

- Any service, pipe, duct, or wire that leaves the building should be carefully insulated and sealed.
- All systems are powered by electricity.
- Where stale air is exchanged for fresh, as much heat as possible should be captured and sent where it is most useful or least detrimental. On cold days, send it inside; on warm days, outside.

There are many fine points to the process, but those are the principles. This is a sensible approach to building a home in this country. Very little energy, which means very little money, will be required to heat and cool these houses throughout their life spans. The owners will be able to operate them without a manual. And if electricity in New York is ever generated in a renewable way, these houses will essentially cease polluting. Call it passive if you like; I think it's pretty badass.

I have no interest in telling people what to build. People come in a remarkable array of shapes, sizes, colors, habits, and tastes. If you can build a house without causing too much irreparable harm, please yourself with its trappings, and not piss off the neighbors, in my book, you have done well.

ART

> *Some lucky lives follow Creation's elegant arcs;*
> *others are Nature's smears.*

The attraction and terror of art, at least for the maker, is that it is unbounded by definition. I'm unable to think of another pursuit in which anything at all might qualify. In recent years, it's not even compulsory that an artist produce an actual thing; an idea or a gesture might suffice. Because I am largely lacking in art education and the theories that sprout around it, I am unable to comment on the purpose or quality of these developments. What appeals to me about art is that I might do entirely as I please, imbue a work with whatever meaning I find important, convey that meaning with the symbols and markings that I hope will be best understood, and blithely ignore whatever the current art hierarchy might have to say about it. I earn my living as a carpenter, so I have the luxury of pursuing art with no concern for its monetary value or societal acceptance. One of the advantages of being unburdened by diplomas is that I lack certification's allegiance to any school of thought. So many meticulously researched theories and scholarly treatises have passed me by unnoticed. I am allowed to think in whatever way I fancy, confident that no scholar's glass will turn its critical focus my way.

In my naïveté, I turn to art for redemption.

A BRIEF IMAGINED HISTORY OF ART

Long ago, in an unknown locale, a group of protohumans gathered near a warm fire of an evening. Their language was young, so stories were brief and practical, lacking the colorations that might hold a listener rapt. One of the group had a wandering mind and a flair for inventiveness. He toyed with a stick, charring its tip, sharpening it to a point against a stone in the manner the entire clan had adopted for hunting, occasionally touching the end to test its warmth. His fingertips became covered in soot, so he cleaned them by transferring the soot in stripes to his forearm. This was interesting, if not necessarily useful. He continued in this vein unobserved, tracing dark stripes on his thighs, triceps, chest, and finally his face. The effect charmed him. He turned to his neighbor, tapping him on the shoulder to get his attention. "Aaaacckkk!" his neighbor cried out; his eyes widened, his ears flared, and his hand grasped blindly for the nearest stone or stick. Now, this was interesting and useful! There was power in that "Aaaacckkk." Soon the entire clan was experimenting with this newly discovered pastime. Berries were smeared on lips, eliciting admiring sighs. Garlands of flowers were strung and hung about necks and limbs in inviting ways. Hunting scenes were scratched into walls, telling tales in the detail their language could not yet manage. Hair was twisted and gathered into mesmerizing shapes. Some even changed their way of walking and gesturing to give it menace, nobility, or sex appeal. The best of these would demonstrate their newly invented movements around the fire while the gathered observers would become so entranced they would pound their hands together or bang the earth with sticks.

Generations passed, and in their inventiveness, the clanspeople found ways to excite in one another a spectacular panoply of impulses, from puerile to pious. Art was born.

Millennia passed. On a warm summer evening, another clan—now humans—gathers at a seaside promontory on Rhode Island's coast. Inside a carefully crafted pile of stones a fire roars beneath the mantel of the great entry hall of the Vanderbilts' "summer cottage." A string ensemble plays gaily; a regally dressed butler announces the guests. An invited couple enters the hall for the first time. "Gadzooks!" the gentleman exclaims at the assembled splendor, his language having become more colorful since that primal "Aaaacckkk." His wife gasps in admiration as Mrs. Vanderbilt descends the grand staircase in her finery. The gears of evolution have made their slow turn.

Since that evening much has gone astray. Great global wars,

famines, and a growing awareness of the world's vast inequities have forced the questioning of our place and purpose. Many have been lost to meaninglessness; so much of our art has become dedicated to its expression. But people have always looked at a troubled world and imagined more perfect things. A medieval tapester sees a horse and weaves a unicorn. An orphaned organist's grandparents survive the Thirty Years' War that the precocious boy might grow to compose the Magnificat. We should not allow ourselves to be swayed into smallness. Laziness and despair need not lead us to the mistaken expectation that meaning is to be provided by existence rather than made in us from its deliberate digestion.

MARKS

I'm fifty-nine years old and I just made my first work of art. Of course, there were heaps of "Helicopters Kill the Dinosaurs" and "Mom at Work" drawings that made their way behind refrigerator magnets. There were music lessons that led to scribbled ditties. There were classes in pottery, life drawing, color theory, and stone sculpture. There was a year at "art" school in Manhattan yielding sheaves of charcoaled smudgings. There were designs for components, furniture, theater sets, musical performances in bars, parties, and fields, and my job for forty years has been to make all manner of things. Finding creative solutions to aesthetic/technical problems is the whole of my work and takes much of my personal time, but I have never called any of it Art.

I'm not trying to draw the distinction between art and craft here; to me there is none. People have wasted time on that discussion for centuries. Highbrows pooh-pooh the place of fashion and fabric arts in their concocted hierarchies, almost entirely because of who makes it and who loves it, but fashion has a far deeper transformative effect on lives and culture than some shock-silly performance artist. Any "craft" can be "art" if its purpose and execution align to make it so. Instead, I am drawing the distinction between the intention behind all the things I have made in my life, no matter how "creative," and the purposeful making of a work of art.

Nine years ago, I received a call from a well-spoken woman who had gotten my number from a longtime colleague. "I'm building a pavilion for a carousel in Brooklyn Bridge Park. We want to make several pieces of custom furniture: ticket kiosks, a manager's desk, benches, and a wardrobe, but several millworkers have looked at it and told us it's too complicated. A man who works for us from time to time told me you might be able to help and gave me your number." These are the kinds of calls that bring most of my paid work these days. "Sounds like you're my customer," I told her.

We met, agreed we were a match, and set things in motion. Many meetings, drawings, and models later, I delivered the furniture pieces to the shining acrylic vitrine of a building she had commissioned. The carousel was halfway installed, its roundabout and much of the mechanism in place when I arrived. My oldest son, Matthew, and I installed our pieces in a few days, watching the completion of the nearly hundred-year-old child's toy with more interest than we took even in our own work. A mechanic who had spent his entire career assembling and repairing carousels supervised. He connected shafts, meshed gears, aligned cams, resuscitated a great calliope, and finally, one bright day, wheeled out the horses and hung them on their stanchions. They were hand-carved wooden masterworks, fiery and prancing in their postures, nostrils snorting, manes flying about, each painted anew in the colors they had worn a century earlier when they made their first ride.

What a grand escapade it was. A sunny knoll, a gleaming building, a merry, whirling carousel—it captured my imagination. I began looking into the history of it, asking my patron and the mechanic to tell me all they knew.

The carousel was fashioned for Idora Park in Youngstown, Ohio.

One hundred thirty years ago, trolley companies began building amusements at the end of their lines to attract customers on their newly allowed weekends off. "Trolley parks" popped up all over the country: Coney Island in Brooklyn, Dorney Park in Allentown, Pennsylvania, and my own Kennywood just outside of Pittsburgh are all remnants of the hundreds of these that were built, each with its own claim to fame. Youngstown had the largest outdoor ballroom in the eastern United States. Often, "claims to fame" need to be qualified a few times to become superlative.

In 1922, Idora got its carousel. Generations of kids clung to those horses, imagining themselves jockeys, cowpokes, knights, and adventurers. In 1984, a fire destroyed large parcels of the park. It never recovered. The carousel was auctioned in parts, every piece won by my patron who lovingly restored it, often with her own hands, until finally it landed at the base of the Brooklyn Bridge, her gift to the borough.

Each time I learned something new about the carousel, Idora Park, or Youngstown and its demise, the story opened a connecting vein from my own history and that of industrial America. Pittsburgh and Youngstown were muscular, dirty, productive machines, driving growth, progress, and the creation of fortunes throughout this country. During World War II, air strikes flattened most of the world's factories, but America was out of reach. Over the course of the war, American industrial records were set and broken; after the war, we were the only country left that had the capacity to make much of anything that the shattered world needed to rebuild.

A myth was born of American industrial might, followed by decades of complacency, lack of upkeep, and a tolerance for inefficiency and waste in the pursuit of immediate profits. When the whole thing toppled in the seventies, steel manufacturers were still operating plants with technologies from the twenties and thirties. The factories of Germany, England, Japan, and later China were shiny, new, efficient, and hungry.

All of this was the backdrop to my childhood, adolescence, and adulthood. My family would celebrate the Fourth of July listening to patriotic anthems blaring from an orchestra barge as fireworks burst gaily over Pittsburgh's three rivers, everywhere rimmed by bloated fish floating belly-up in bunches. Cinder-belching steel mills lined the banks for miles in each direction making fireworks of their own every night of the week. Today they are gone.

For several years, this collage of images, with the carousel whirling at its center, nagged at me. Maybe I could write a book. I gave it a few passes; nothing seemed to say what I wanted. "I'm not a writer," I admitted aloud. I knew I wanted to express something about all this; I just didn't know exactly what or how. So I enlisted my son Martin and his girlfriend Lila to help me with what I kept calling "the Youngstown book." We loaded two cars and went to Youngstown for a week to see what we could find.

Youngstown is less remarkable for what is there than for what is gone. All that's left of Idora Park is a few concrete pads with bolts protruding where the parallel tracks of the two roller coasters were once anchored. A sweeping curb traces the path of the long since empty midway. The abandoned acres are fenced in and ostensibly off-limits, but no one is there to keep the curious out.

Along the river, the steel mills' towering furnaces were long ago blown from their perches and hauled away for scrap. A few warehouses remain, some with businesses that moved in like hermit crabs, making a go of the old shell. A few miles out of town, there is a still-standing mill where it looks like everyone got up and walked out on the same day. A fence surrounds it, and signs are posted about, but anyone who wants to can walk right in, climb on trucks with enormous wheels, and even rummage through paperwork scattered about its offices, some of it still weighted by coffee mugs lined with dried brown rings. One day, in the not so distant past, everything just stopped.

Plenty has been written about how this happened. Towns throughout America suffered similar fates: Local factories closed; industry moved overseas; jobs dried up. Half of the country was left with little worthwhile to do. Enterprising companies—ones that reinvested in their factories, owned their equipment and property, concentrated their energies on efficiency and progress— were undone by the corporate raiders of the eighties. Financially outflanked, they were sold off for parts by some of the very people for whom I have built homes. What is anyone supposed to say about all this? It's too unfathomable. I was stymied; no book was forthcoming.

Martin, Lila, and I came back to New York with several rolls of used film and not much of a plan. I took a few more stabs at writing; it all fell flat. Between efforts, I would retreat to my couch and practice my guitar as I have nearly every day for forty-seven years. Music had long taken a back seat in my life, but at fifty the exodus of my children made more room in my days. After a few years of concentrated practice, I finally started liking my own playing. A close friend and I formed a jaunty local ensemble, and unexpectedly, I began writing songs at a surprising clip. Songs could always do for me what books never could: focus on one or two ideas, talk about them in oblique yet highly personal terms, and wrap things up before anyone noticed it was me who made them cry.

A few songs began to take shape around the seeds that the carousel and Youngstown had planted in me. For the first time in my life, I had specific things I wanted to say, things I considered meaningful and important, people I wanted to have hear them, and a way of getting my point across. One song followed another, each with a different setting and point of view, until I had a dozen. I sent a few of these off to a producer/engineer/drummer I had met, worked with on occasion, and didn't know well, but whom I could trust to deliver an objective opinion. He said we should make an album together.

Several months later, we began. All those years of effort have made me into a pretty okay guitar player, a serviceable singer, and a semi-enthusiastic performer. Given my limitations, I knew well enough that I should practice the songs I'd written until they were stage-ready. My friend and bass player, Rob, was more than happy to help, lured not only by the prospect of playing in a genuine professional recording studio, but also by his oft-repeated opinion that the project was worthwhile and the songs were good. We worked regularly and diligently through twenty Sundays until the day came when we loaded my little pickup with instruments and drove to a basement studio in Queens, The Madhouse, helmed by Mark Ambrosino, its owner.

Mark is the most gracious musician I have ever met. His drumming is unmatched, made virtuosic by his uncanny ability to highlight the transitions between passages and emotions, and grounded by a backbeat so religious that every ring of the snare is a relief. He carries his history quietly, but after a few late-running workdays, he'd let out a tale or two from his long career of playing with the greats. For three months, we spent hours together, swapping stories, recording parts played by a parade of professional musicians far more accomplished than I am. We developed a mutual respect that's rare and infinitely valuable. Whenever my confidence flagged, he would prop me up with a pep talk, refill the coffeepot, and start us back in on the business at hand. Decades of work in music, and tireless devotion to its making, have made him strong, humble, and forthright. Each night I would leave worn out from concentrating, having exercised amateurish muscles to capacity. He would carry on for hours after I left, editing, arranging, and compiling the best of the tracks in preparation for the next day. It was exhausting and blissful.

The work of thirteen people went into crafting that album. It's called *Miles of Dirt*. There's not a line in it that doesn't have multiple meanings for me. It is quirky, dense, and personal, while drawing on all the sounds, rhythms, and lyrics I loved growing up. It certainly has passages that sound a little Johnny Cash or Isaac Hayes, but no one would mistake it for an homage. Martin and Lila are making drawings and photographs to accompany the album, all of which will be hand bound into six volumes, one to keep and five to give and sell. A local letter-press/bookbindery is hand printing the copy and sewing them in an outdated fashion that suits them perfectly. I'm making special cases and stands to hold them all. Perhaps someday we'll take the conventional route and put it all on the internet for streaming, but not now. It's an artwork, the only one I've ever made, the product of a lifetime of effort and pondering. Let it be "content" some other day.

———

Recently, I bought an old firehouse in Newburgh, New York. It was the home of the Columbian Hose Co. No. 2. Everyone in the neighborhood knows this, not because Newburgh is a town of history buffs, but because it's carved in granite above the door. I've only been there a handful of times and I've already heard three adolescent jokes about the name; at least the neighbors are paying attention.

Newburgh is a town to which people from the nicer places in the Hudson Valley move when they get divorced. It's run-down, a little dangerous, and brimming with possibilities, a lovely place to make a new start. Rebirth feels like more than an idea here, and it feels like more than an idea in me.

This town had a good run. For decades it grew, thrived, built, expanded, and was undone by the same forces that tore apart so many of America's industrial cities. Towns like Newburgh might seem lost, but they have one major advantage over the rich, flourishing cities of the world: affordability. Fabricators, painters, sculptors, dancers, actors, all need space; few find themselves well funded.

In one forgotten neighborhood after another in New York, for the last forty years, I've sat aside as intrepid artists found lofts and studios where they could do their work. Then a different wave would come, of architects, designers, shop owners, restaurateurs. In a decade or two, the neighborhood would be transformed. The artists and fabricators would find themselves priced out, and off they'd go to the next affordable place, usually a little farther out of town. This is called alternately gentrification and urban renewal. In either case, it generally happens with no regard for the people who lived in the area before the artists arrived. This is an opportunity wasted.

I was given a ride recently by a young cabdriver, a Newburgh native, who could see from the ground the changes that are coming to this town. In his off-hours, he is organizing a training center for interested friends and neighbors, enlisting local contractors and tradespeople to create an apprenticeship program so that those who live here might benefit from the development that is coming. Rebuilding a city takes an army of workers. How much better it would be if they could be people who grew up here, people who, with a skill, a market, and some hard work, could buy a home in their own town, fix it up, and derive the same benefits that the new arrivals hope to find. I should have been so smart and farseeing when I was twenty-three. There will be room in my workshop for a trainee.

I have worked in construction for forty years. In those years, I've built more things and more interesting things than I ever thought I would. I don't know how many more years of building I have left in me; I'm hoping for another twenty or thirty. But I would like to change my focus. I want my new shop in the firehouse to be a "Workshop of Wonders" where, for the first time in my life, I can build the things that have been brewing in my imagination all these years: grand things, personal things, things that may have meaning only to me, but that might feed the imaginations of others.

The freedom to spend my days making whatever I want is a fearsome proposition for me. The last forty years have been driven by necessity. Thirty years ago, given the opportunity to do whatever I wanted with my days, I might have done nothing at all. I don't think I'm capable of that anymore. Necessity has ground the habit of industry into my bones. A long weekend at the beach is more than I can bear and I still feel like I haven't worked if I don't touch a tool all day. I'm less afraid that freedom will expose a lack of inspiration than I am that I will be unsatisfied with the results of my creative efforts. After all these years, I feel like a beginner again.

Mark Ambrosino and I have agreed to record a second album. This time it will be a collection of recordings like other musicians make, and this time we will send it into the wider world. The songs are already written and practiced. By the time anyone can read this, it will probably be done. If anyone wants to listen, it will be called *Hard to Tame*. It's much less likely that I will be found performing in public with any frequency.

Under my workbench, I have several crates of guitar parts that I started making six years ago. I'll begin with those. Half-made things are like half-kept promises; they gnaw at my conscience. I was so sure when I started that I could make guitars as lovely as any I own. Now that the parts are covered in a few years' dust I'm less convinced. Luthiers spend decades perfecting their craft, and I own a few really fantastic guitars, so the stakes are high. There's only one way to find out.

Then there is the series of beds I designed and never built. Years ago, while watching a movie whose title I forget, I became enamored of Chinese wedding beds. I love that one can climb right inside them and be enveloped in their romance and mystery. Since then, I've come up with six distinct and varied designs inspired by the idea. They are less furniture than environments: One is shimmery, filled with shell-like glass disks arranged in a canopy; another is tendinous and embracing like the interwoven vines in an ancient bog; a third will be made with fibers, fabric, and thread, incorporating weaving, embroidery, lashing, and knotwork. The rest are just as far-fetched and ambitious.

Finishing the guitars and beds will take care of the first few years. With some luck they might satisfy my nagging desire to build, or at least attempt to build, a few perfect things. Both projects are of a scale where I think it is possible to get every detail right. Even if I'm misguided, to me it's a wildly attractive challenge.

Once all of that is under my belt, who knows what I will undertake. I may want to draw inward and work on small and simple pieces, or I may cut loose and build things that crawl up the walls and cover the ceiling.

I can't imagine limiting myself by scale or scope. If a project comes along that's grand enough and tantalizing enough, there's little that would keep me from assembling a crew and taking it on. There are still plenty of puzzles in the building world that I can be tempted to tackle, and I'm still not the solitary artist type.

It hardly matters what I make. What matters is the doing. I have finally reached a place where I have the chance to live up to my own expectations. I can't very well step to the side and let someone else volunteer for the task. Fear be damned. I've already ordered the sign. It says WORKSHOP OF WONDERS.

Epilogue

I am mistaken.

Two years ago, I was pedaling my folding bicycle up Park Avenue to the Harlem train station on my way home from work. A steady rain slicked the streets and blurred the evening landscape. I splashed my way through rivulets and puddles, riding gingerly as my history had taught me to do. Twenty months earlier, I had been caught between a Cadillac and parked cars when the driver pulled over without checking the bike lane. The fall fractured my left wrist. I was on my way to my band's musical performance, so I let shock cover the pain and played the show anyway, finally seeing a doctor the next day.

Ten months after that collision, I crashed into a runner near the Metropolitan Museum of Art who darted from behind a bus. The first thing to hit the ground was my right hip. I heard it break but didn't want to believe it, so I rode the rest of the way to work. I took three Advils and spent an hour hobbling around the jobsite before Pauline insisted on taking me to the emergency room. That evening, a surgeon installed three four-inch pins that reattached the ball of my hip and I learned why people find fentanyl so appealing.

Those two accidents were my fault. I had the nasty habit of riding toward cars that were driving thoughtlessly. The impulse to teach the Cadillac owner a lesson backfired. The runner I hit didn't look where he was going; he didn't have to—it was his light.

Broken bones have memories of their own. On the night of this story, my legs cranked carefully up Park Avenue; I kept close to the edge of the street, letting my two harsh lessons guide my way. I steered my bike across a three-foot puddle. Its shiny surface concealed a gaping pothole, deep enough to swallow my front tire. My upper lip broke my fall, halting the momentum of my entire body's weight against the jagged rim of the crater. A second later I pulled myself up by the curb. Blood covered the front of my shirt. My tongue probed for my front teeth and found them flattened against the roof of my mouth. A woman in a Honda pulled over crying, "Oh God, I'm so sorry," and handed me a handkerchief. She had been nowhere near me and bore no responsibility for my distress, which made her reaction all the more alarming. Half a block away, a young doctor in scrubs was waiting to cross the street. He saw me hustling toward him and turned my way. "Is it bad?" I asked, lowering the handkerchief. "Uh, yeah. There's an emergency room right over there."

I folded my bike and walked the block and a half to the emergency room. The triage nurse chastised me for bleeding on his desk. He had me fill out a handful of forms and led me back to the waiting area where, for the umpteenth time, I sat and thought about my careless existence. A tender resident wired and sewed me up as best he could and sent me on my way. A few months later, my dentist screwed two new teeth into the spot where he had induced fresh bone to grow over the cracks in my upper jaw.

I don't miss my old teeth. I had already knocked the same pair out as a preschooler when I slid off the back of a teeter-totter wanting to see what would happen when my unsuspecting counterweight of a friend crashed to the ground. Of the missing adult teeth, one had long been dead and gray, victim of a left hook I never saw coming. The new ones are shiny and white, far better-looking, even if it still feels like I've got someone else's teeth in my mouth.

Falling on my face hurts, but eventually it gets better. Nothing ever heals all the way. I carry all my life's pain in my gait, in the fear that flashes behind my eyes when someone comes too close, in the lies I tell to cover my shame, and in the memory of the suffering I have inflicted on others that can't be undone.

Every well-intentioned teacher I have had taught the same lesson: Do everything as perfectly as you can. One hundred percent correct is the only goal worth pursuing. It's a lesson that demands rigor and promises the highest reward. There's nothing wrong with the lesson. How could there be? It's the recipe for perfection. But the harder lesson to learn is that failure, brokenness, weakness, and error cannot become the targets of our derision. They are our constant companions, and the most difficult to welcome.

There is so much that is broken in this world. Most of what we do is not done well, and nothing is done perfectly. Grave mistakes have been made. Our planet is a great cacophonous, limping mess. I wish that I could say it was not my fault, but I find that I cannot, not when I'm honest with myself. We have built for ourselves exactly the world we deserve. It is a precise reflection of how far mankind has come, in every horrid and glorious detail. I would love to shout, "I am a student of beauty, honor, and all that is good!" I would delight in exempting myself from the worst of everything. But I cannot, not while I am still so steeped in transgression.

I have not done much good in this world. By now, three-quarters of what I have made decorates the city dump. I have been helpful here and there, but harmful in equal measure. Much of what I believed has turned out to be nonsense, even those things I held dear. It has been my terror of being tested that has kept those beliefs so firmly in place. Comfort and safety are strong opiates against the disconsonance and fear they stave off.

I wish I could offer aphorisms that set myself and others happily on our way. They're easy enough to find. Every Disney movie exhorts each of us to follow our hearts. I'm not sure old Walt understood what the human heart contains. I doubt he even noticed that while Snow White gaily followed her heart, the Evil Queen just as eagerly followed hers.

Joseph Campbell said, "Follow your bliss." Few remember that he spent his entire life seeking the sources of bliss and tending them. He didn't find bliss on a refrigerator magnet in his kitchen. Bliss rarely comes unbidden. Most of us are ill-prepared for ecstasy.

Bliss and ecstasy are real. At times they envelop us, perhaps all the time, although we notice them only rarely, wrapped up as we are in our mutterings over the slights we have endured. We are hard nuts who resent being cracked. It hurts, and disturbs our sleep.

Up until now, I have resisted telling the morals to my stories. They're in there, but it's far more fruitful if you find them yourself. As this is my final opportunity, I aim to be as clear as I can, because this idea is a hard thing to hear.

THE MORAL OF THE STORY

Every error is a door.
The keys to our cells are hidden behind our mistakes.

The first accident in my story was entirely my fault. I wanted to punish the Cadillac driver for his inattention and carelessness. I paid with my wrist.

The second accident was entirely my fault. I ran a red light, as most bicyclists in New York mistakenly feel they are entitled to do. I couldn't see around the bus. The runner leapt innocently in front of me, and I clobbered him as I sped. He got up and ran away with only a curse, but I could have injured him as badly as I did myself.

By the third accident, I had learned my lessons and curbed my mean and reckless impulses. I was exercising caution, fully aware of the hazardous conditions in which I rode. And yet, it was still entirely my fault. I could have walked the bike to the train station. There was no hurry. I don't even mind getting wet.

I am big and strong; I like to think of myself as tougher than others, more thoughtful, more determined. But I am as fragile as anyone. I can be broken, and in the end I surely will be. My upper lip was torn to the nostril, my teeth were destroyed, and my brain was injured enough that for six happy weeks it was incapable of producing the phenomenon called worry.

I am mistaken. Anger, Malice, Indifference, and Superiority

lurk behind everything I do. When they are absent, Cowardice, Envy, Avarice, and Spite are happy to take their place. At every turn, they steer my intentions, sometimes with a nudge, sometimes with a shove. They exercise their wills in us all.

Look at the world we live in. Look at the beauty and the savagery of this place. It is an exact model of the human psyche. How else could it have become what it is? We made it, and it made us. Everyone we encounter is beset by grief and shame and sorrow. We are all buoyed by hope and wish and admiration.

There is a great deal in the world that inspires me. There are good ideas out there, things we can be proud of. They are to be cherished and preserved.

But the great secret to freedom is to look as unflinchingly as we can bear to at the myriad ways in which we are wrong, to relinquish the foolish thoughts and beliefs that force our actions into repetitive cycles of suffering.

How else can I find the courage to live as I long to? How else can I understand the pain others live by? How else can I repair a shred of this damaged world?

Don't be afraid to be wrong. You are. I am.

Work hard. Love easily. I will do the same.

FINALE, FOR NOW

 People are remarkably similar. We have the same organs, limbs, and facial features. Our brains work by the same mechanisms, using the same proteins and energies. We don't differ by content, but by degree. As much as anything, chance has made me what I am. The few things in life I have learned a little bit about came to me because they were close at hand, so I picked them up. Someone on the other side of the Earth grew to adulthood in different surroundings, with different influences. We learned to think in divergent ways, and to believe that those differences are important, even vital or sacred. Are they?

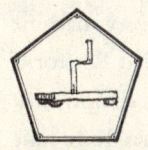 History is long and the world is vast. You'll never get to all of it. It's hard to know what to do. But even in my childhood home, where I lived for only nine years, there were a few things that captured my attention again and again. The dining room was wallpapered in scenes of what I always imagined to be Horse and Buggy Pittsburgh. Our fireplace had a cast-iron and ceramic gas burner that came alight with the turn of a valve and a ready match. In the attic, my mother's Swedish loom worked magically with shuttle and treadles and heddles. A table saw whirred its solid song in the basement. And then there was that piano, like a landscape, from its ribbed underbelly to its iron and spruce soul. Your world can tell you what to want.

 What we do makes what we become. Working at something, or a few things every day, with guidance and attention, changes our physiology, psychology, and our feelings about ourselves and the world. Accom-

plishment and pride walk hand in hand. With daily effort, a student might only add 5 percent of new material in a month of practice. But, just like the magic of compound interest, in thirteen months, that adds up to just shy of 100 percent. If the same student carries on for forty years, they will no longer be the same student, not even close. To paraphrase Herb Alpert's trumpet teacher, "It's not the horn, Herb; it's you."

 Neuroscientists once posited that Precision and Expression inhabit opposite sides of the brain. This seemed deeply important at the time. For some reason, few neurologists or psychologists back then thought much about the robust connection the corpus callosum provides between the brain's hemispheres. Three hundred million nerves can do a lot of talking. An electrical wire of the same gauge could power every machine in my workshop simultaneously. Anyone who has studied Rhetoric, Drawing, Fighting, Poetry, Music, Dance, and even Science knows that Language and Math are happily married. Every pursuit has Technique and Eloquence, Exactitude and Phrasing, Truth and Beauty.

 For the longest time, I thought life would eventually make sense. It's divinely simple; it's devilishly complex. It's ruled by natural laws; it's the very definition of chaos. We are blessed; we are damned. The universe is within me; I am entirely alone.

Some people say this world is devoid of meaning, and yet my heart beats with such purpose.

 People used to speak differently. Until I got used to it, Shakespeare was a puzzle; Chaucer still is. Not so long ago, when a person got a notion to do something, they would say, "I mean to do this" or "I mean to do that."

In the last few years, I've taken up the phrase when I set my sights on a goal. I like its varied and sage implications. "I mean to get this done" says a lot about my intentions. There's a recognition of the need for focus, the likelihood of struggle, and a hopeful appeal that Fate might allow the thing I wish for. It doesn't guarantee success; I could succumb to weakness and sabotage my own efforts; I may be reaching beyond my abilities; circumstances might make my wish impossible, but at least I can say out loud where I stand.

 Most of life is work. We live on a planet where everything needs tending: businesses, relationships, homes, gardens, machines, interests, bodies, psyches. Everything here falls apart unless someone makes an effort to keep it going or improve it. And things *still* fall apart sooner or later. Pleasure can be found in doing things we enjoy, but there are so many things that need to be done whether we want to or not. It is a considerable achievement to reach the stage in which most of the things in your sphere are well tended and working smoothly. Each time an endeavor reaches that stage, it is cause for pride, even celebration. One need not throw a party, although that can be fun. Sometimes it's enough to stand with a neighbor across the fence and listen when they tell you, "What a beautiful garden!" Then you can say something pleasant and true, like, "Thank you, I'm very proud of it."

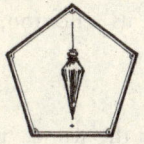 The only standards that matter are the ones to which I hold myself. It can be seen as a kindness that others hold us to standards that we haven't yet learned. If they hold themselves accountable in the same degree, they are to be admired. If not, they aren't doing themselves any favors, even if they might be doing us one. The higher a standard I'm able to set for myself and consistently work toward, the more challenging life becomes, and the more frequently I meet my shortcomings. Then the

definition of tolerance shifts, and I have to find a way to live with myself. Acceptance is what modern counselors tell us to strive for when faced with our frailties. I have never gotten that far. Tolerance I can manage.

 As a child, when I fell and hurt myself, the most I could hope for was a cursory kiss and a handkerchief to wipe away a tear or two. If no adult was around, tears were hardly worth the effort, unless the pain was truly remarkable. Children are easily taught to make victims of themselves, to savor self-pity over shame or remorse or determination. Certainly, it's fair to offer solace when they suffer. Affliction is just as worthy of acknowledgment as achievement. But a skinned knee is not a ruptured spleen, and a scribble is not a masterwork. Sometimes "You'll be fine, stand up" or "How nice, dear" is sufficient and leaves room for growth.

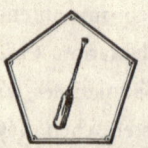 I do not know a greater pain than the loss of a companion. Friends, lovers, family, they all leave us in time. Pain and love live side by side in my heart. They awaken one another. I have been inconsolable in calamity and euphoric in connection. Everyone we walk among lives between devastation and joy. We all lose everything. In our affliction and elation, we have a choice: Will I exact revenge for the sorrow the world has brought me, or will I carry that sorrow, see its reflection, and let it goad me to compassion?

 We are given the opportunity to rebuild the world. If you want to try, start small. If you succeed, expand.

ACKNOWLEDGMENTS

If it weren't for Lauren Sharp, my agent at Aevitas Creative Management, this book would not exist. On December 8, 2020, I received her first email. She had pried my contact information out of Burkhard Bilger, a longtime staff writer for *The New Yorker,* who had recently published a lengthy profile on my work. Plucking people from obscurity is a sort of specialty of his. Lauren and Burkhard had conferred and decided that I should write a memoir. Lauren's email asked if I was on board. I am an incessant storyteller, and this wasn't the first time I had been told to write things down. The fly in the ointment was that I had never recorded anything aside from a few twangy songs. There were no pictures, no notes, no journal, not even respectable financial records. Who would think that if I put fingertips to laptop keys that anything worth reading would come about?

Answer: Lauren Sharp

Lauren has a natural trustworthiness, borne out by every interaction I have had with her over these two and a half years. She also has the persuasive powers of an experienced kindergarten teacher, which is apparently just what a curmudgeonly character like me needs. I don't have enough fingers to count the number of times I told her no, and yet I found myself sending her story after story. What could be more encouraging than hearing, "These are beautiful. Write a hundred more and you'll have a book!"? I'll be damned if I didn't do just that.

Thank you, Lauren. It's been one weird, unexpected road you've led me down.

Lauren sent the book proposal off to various publishers. My first ever Zoom calls were arranged. People said nice things that surprised me. Of all the calls, Random House's team was the most convincing in their enthusiasm. Mark Warren, my now editor, had clearly read the darn thing and had been moved. Three of his colleagues were on that call, publisher Andy Ward, editor in chief Robin Desser, and deputy publisher Tom Perry, and all of them had something compelling to say.

A few weeks later, Random House became my publisher and Mark Warren became my editor.

Since that day, Lauren's colleague Erin Files has busily sold this book to the wider world. I thank her for her perspicacity and diligence.

Publishing is a world about which I know nearly nothing. Writing is something I have avoided since grammar school. If it weren't for Mark Warren, what you have in your hands would be a paella of disparate, unconnected tales from the trade. Because of him, it is a book.

Thank you, Mark. I couldn't have done it without you.

My son, Martin Ellison, drew each chapter's frontispiece, and several of the stories' illustrations. I could have hired a veteran illustrator for this project, but I don't think I would have gotten a better result. Martin was twenty-six when he started. He has kept drawing images for this book for more than a year. His hand is so sure, and his images are so strong. He made this book more than it ever would have been without him.

Thank you, Martin. Your work is beautiful.

Thank you to all of the people at Random House who take scribbles and key taps and turn them into the bound, decorated, word-and-picture filled thing that people have turned to for centuries to find inspiration and meaning. Thank you to production editor Ada Yonenaka for taking such great care with these sentences, and to designer Greg Mollica for your collaborative spirit and the beautiful cover, assistant editor Chayenne Skeete for all your help throughout this process, and publicist London King for connecting this first-time author to the world.

Finally, thank you to the special people in my life who have loved me, tolerated me, and shown me understanding. It has meant the world.